Letter to a Young Farmer

Other Books by Gene Logsdon

FICTION:

Pope Mary and the Church of Almighty Good Food

The Last of the Husbandmen

The Lords of Folly

The Man Who Created Paradise—a fable

NONFICTION:

Gene Everlasting

A Sanctuary of Trees

Holy Shit

Small-Scale Grain Raising

The Mother of All Arts

All Flesh Is Grass

The Pond Lovers

Good Spirits: A New Look at Ol' Demon Alcohol

You Can Go Home Again

The Contrary Farmer's Invitation to Gardening

Living at Nature's Pace

The Contrary Farmer

The Low-Maintenance House

Money-Saving Secrets: A Treasury of Salvaging
Bargaining Recycling & Scavenging

Gene Logsdon's Practical Skills

Wildlife in the Garden

Organic Orcharding: A Grove of Trees to Live In

Getting Food from Water: A Guide to Backyard Aquaculture

Trees for the Yard, Orchard, and Woodlot

The Gardener's Guide to Better Soil

Successful Berry Growing

Homesteading: How to Find
New Independence on the Land

Two-Acre Eden

Wyeth People: A Portrait of Andrew Wyeth
as Seen by His Friends and Neighbors

Letter to a Young Farmer

How to Live Richly without Wealth
on the New Garden Farm

GENE LOGSDON

Foreword by
WENDELL BERRY

Chelsea Green Publishing
White River Junction, Vermont

Cover illustrations by Vladimir Yudin / 123RF Stock Photo (*top*) and istock.com/
blue67sign (*bottom*)

Project Manager: Alexander Bullett
Project Editor: Benjamin Watson
Copy Editor: Angela Boyle
Proofreader: Helen Walden
Designer: Melissa Jacobson
Page Composition: Abrah Griggs

Printed in the United States of America.
First printing February, 2017.
10 9 8 7 6 5 4 3 2 1   17 18 19 20 21

Our Commitment to Green Publishing
Chelsea Green sees publishing as a tool for cultural change and ecological stewardship. We
strive to align our book manufacturing practices with our editorial mission and to reduce
the impact of our business enterprise in the environment. We print our books and catalogs
on chlorine-free recycled paper, using vegetable-based inks whenever possible. This book
may cost slightly more because it was printed on paper that contains recycled fiber, and we
hope you'll agree that it's worth it. Chelsea Green is a member of the Green Press Initiative (www.greenpressinitiative.org), a nonprofit coalition of publishers, manufacturers, and
authors working to protect the world's endangered forests and conserve natural resources.
Letter to a Young Farmer was printed on paper supplied by Thomson-Shore that contains
100% postconsumer recycled fiber.

Library of Congress Cataloging-in-Publication Data
Names: Logsdon, Gene, author.
Title: Letter to a young farmer : how to live richly without wealth on the
    new garden farm / Gene Logsdon.
Other titles: How to live richly without wealth on the new garden farm
Description: White River Junction, Vermont : Chelsea Green Publishing, [2017]
Identifiers: LCCN 2016050357| ISBN 9781603587259 (hardcover) | ISBN
    9781603587266 (ebook)
Subjects: LCSH: Agriculture--United States--Anecdotes. | Farm life--United
    States--Anecdotes.
Classification: LCC S521.5.A2 L64 2017 | DDC 338.10973--dc23
LC record available at https://lccn.loc.gov/2016050357

Chelsea Green Publishing
85 North Main Street, Suite 120
White River Junction, VT 05001
(802) 295-6300
www.chelseagreen.com

In honor of my friends,
Wendell Berry, Wes Jackson, David Kline, and Maury Telleen,
whose words and actions have so much influenced
the blossoming of the garden farm.

"In the future, most farms will be just very big gardens."

ROBERT RODALE,
at the time publisher of Rodale Press and editor-in-chief of
Organic Gardening and Farming *magazine,*
talking with the author in 1978

CONTENTS

FOREWORD

etter to a Young Farmer is Gene Logsdon's valedictory statement, written during what he knew was his final illness, finished at virtually the last minute of his working life. To his friends and devoted readers, those circumstances alone would make this a remarkable book and, moreover, a triumph in keeping with his character.

But this book is as variously remarkable as Gene himself was. By inheritance and conviction he was a holdout, an agrarian. He was, to use one of his favorite terms, a "contrary farmer." By that name he distinguished himself and others of like mind from virtually all the kinds of modern ambition and success. The only diploma I have on my wall is a printed certificate naming me a member of The Contrary Farmers of America. It is signed by Gene.

What are contrary farmers contrary to? By the testimony of this book, they are contrary to getting "big" at any cost, to buying everything new and expensive that is recommended by experts, to dissolute economic and social behavior, to the fanatical pursuit of "more" and "better," and to farming as territorial aggression or surface mining. This is a contrariness that has no doubt of its proper Earthly place or of the validity of its cultural tradition.

Among the ranks of the liberals and conservatives, you will find no contrary farmers hot-eyed and chanting slogans. Contrary farmers at times have had political representation. Some have at times been politicians. But this book is unlikely to be read at present, much less quoted, by any candidate for national office. Contrary farmers now are represented by no party. But they *are* a party—or a party of sorts—to the extent that a good many of them still exist, are likely to recognize one another, have much to say to one another, and by nature or second nature are resistant to sales talk and expert advice.

And so, perfectly in keeping with the character of its author, this is a radical book. It is quietly, bravely, intelligently, amiably, and gleefully radical, intended, as Gene wrote in the Preface,

> *as fingers-crossed recognition of the decline of the Industrial Revolution, at least as it relates to farming, and the rise of a new decentralized agriculture based on home economy, not the so-called "cheap food" economy. . . . I mean to celebrate the rise of the smaller-scale, bio-intensive, environmentally friendly garden farm, a place where food quantity and food quality merge to bring about food sufficiency guaranteed by human self-sufficiency.*

Gene is proposing, in short, to replace "get big or get out," the "conservative" slogan of Secretary of Agriculture Ezra Taft Benson in the 1950s, with another slogan that is truly conservative and far more democratic: "get small and stay in." Beyond that, the opportunity for contrariness becomes pretty scarce.

This is also an eminently generous book, conceived and understood as part of the long conversation about farming, which has no beginning that we know and no end that we can foresee. In it an old farmer is talking to a young farmer, who will become an old farmer talking to a young farmer. This conversation is not just an interesting artifact or a good idea. It obeys the law of human culture, on which the life of the Earth depends, as it depends on the laws of nature.

In the role and responsibility of an old farmer, Gene is talking, to any young farmer who may be listening, about what he has learned, what he has in mind. There is something here of an old man's pleasure in listening to himself as he discovers word by word his ability to *say* what he knows and what he thinks. And almost every page of this book conveys the pleasure Gene took in the subject of farming. His interest in it and in all the details of it was unending. As he talked of it, he clearly was having a good time. He was telling what he had learned from history, from his family's history, from his reading, from his experience.

Most of all, I think, Gene is telling what he has learned from the all-but-perfect twenty-acre farm that he and Carol, his wife, made by living on it and from it for four decades. I visited that farm a good many times over the years, and I can say for it that when I was there I was always happy. I was happy of course because of our friendship, companionship, and conversation and because of the incomparable bounty and goodness of Carol's cooking. But I was made happy also by the place itself, by the artistry and relish apparent in all of its details, and because it was so obviously the focus of the large, complex love for it, for the farming of it, for the dwelling in it, that Gene and Carol shared with each other and with their family and friends.

What at last seems most remarkable about this book, and what moves me most, is the tone of it, which has from beginning to end the necessary calm of sense-making. Gene stays aware that he is writing a letter to a young farmer, a letter not meant as the last word but as his part in an ever-continuing conversation. His wish is to be encouraging and to be useful. In the midst of our epidemic fear of the future and its so-far predicted emergencies and catastrophes, here is Gene patiently, quietly, with the right touch of merriment, talking about the small, really possible ways of solving our one great problem: how to live on the Earth without destroying it. He is talking about a profound change in our use of the land, which, with all it implies, cannot come as a political or technological "solution." He is asking instead for what he calls "at-home farming," which can come, for it is already coming, from human culture and human love.

As this book shows, Gene is sufficiently aware of climate change, which is presently the most-favored catastrophe. But as a contrary farmer who can see and think, he is also aware that land and water are as much endangered as the climate, and that the two dangers are really just one: Our ways of degrading the climate are in no way distinguishable from our ways of abusing the land. Because he knows this, he speaks exactly at the intersection of economy and ecology, where the world will or will not be made better.

Well. Bless his heart.

Wendell Berry

PREFACE

I f I were to write a book about softball, my purpose would
be to describe the fun of the game and the fine points that
can mean winning more often than losing. If readers went on
to work their way upward into championship tournaments—even
started making money at it—I would be pleased, of course. I would
brag to one and all that my book had helped. But the purpose of my
writing would be just to describe the pleasures of playing the game.

And so, in the same spirit, I offer this book about farming, as
a sort of open letter to everyone who likes to grow and eat good
food, especially to younger garden and pasture farmers and their
customers, who are now leading us into the new post-industrial era
of agriculture. It is mostly about how working a farm can be even
more exciting and enjoyable than playing softball. If you can use
what I write to turn your farming into a money-making proposi-
tion, I will strut around and pretend that my book helped. But my
main purpose is to teach a love of the art and science of growing
good food, plus the personal contentment that can follow. And
especially, if more people decide to live this way, how much better
off the world would be.

I also intend the book as a fingers-crossed recognition of the
decline of the Industrial Revolution, at least as it relates to farming, and the rise of a new, decentralized agriculture based on home
economy, not the so-called "cheap food" economy. With a mixture
of "how to" and "how great" examples, I mean to celebrate the rise
of the smaller-scale, bio-intensive, environmentally friendly garden
farm, a place where food quantity and food quality merge to bring

about food sufficiency guaranteed by human self-sufficiency. There is no name yet for this new economic model. Through the spreading pastoral and garden paradise that results from it, a new kind of farmer, also without a proper name yet, dances tippy-toed in the embrace of nature, abandoning a century of trying to "get big or get out." Perhaps our old cultural motto of "root, hog, or die" will be replaced by "scratch, hen, and live." Or "get small and stay in."

CHAPTER 1

~~~

No Such Thing as
"The American Farmer*"*

For years the prevailing mantra in farming has been "get big or get out." With that challenge, there emerged an image of the farmer as a bold, derring-do worshipper of big business, willing to take huge monetary risks to expand acreage and stay alive in farming. He was praised for doing so because (so the wise men of economics all said) expansion was the only way to survive. But because whenever one farm increased in size, another went out of business, it would have been just as accurate to say that expansion was the surest way to fail. Unnoticed in the economic battle were farmers who were content to "stay small and stay in."

However accurate or inaccurate those farmer images might be, they don't reflect what is going on now, as people with all sorts of backgrounds are getting into the food farming action. These newcomers are a cross between gardeners and traditional farmers, practicing what I like to call *garden farming*: growing plants and animals on a small scale, as much for pleasure as for the ability to make money at it. Some older farmers like to refer to this new segment of society, with a bit of a curl in their lips, as "city farmers," and indeed they are more correct than they imagine. The impetus for garden farming is coming more from urban society than from the backcountry where the tall corn grows. It well may be that most food producers in the future will be city farmers.

Way back in 1929, Wheeler McMillen, longtime editor of *Farm Journal* magazine (as a boy, I loved his writing style), published a book, *Too Many Farmers*, that tried to address the knotty problem of how to make a biological process like farming work as a profitable industrial business. The book was resurrected by the National Farmers Organization in the 1960s and roundly condemned as the voice of profiteering agribusiness interests. Actually, it was a rather fair-minded treatment of the subject, I thought. I interviewed Wheeler in 1982 for *New Farm* magazine (February issue) when he was eighty-nine years old and asked him if he still thought there were too many farmers. Since only a fraction of the farm population of 1929 remained and there were still "too many" farmers by Wheeler's reasoning, the question was embarrassing, but it did not seem to disconcert him. "Today, I'd say there are too many acres being farmed," he replied. Although he defended himself ably and had good words to say about organic and natural farming, which he went on to predict would become the force that they are today, he seemed unwilling to look at farming as anything other than a money-making business. He would never have thought to write a book titled *Too Many Gardeners*. But, as I questioned him sharply, he did end up objecting a bit to my calling him a hard-headed businessman. "We have all been guilty of over-emphasizing financial success," he said. He then went on to say, somewhat gratuitously, I thought, but with the spark of humor that was always in his writing, that "when a family lives in honest self-respect and raises children to decent ideals, who is to object if some loafing and fishing results in their piling up less loaves and fishes." He could not quite pursue that line of thought any further, however, or consider the idea that if farming were only an industrial, for-profit enterprise, there would always be too many farmers until only one farmer was left.

Whenever I hear a commentator or politician (or sometimes even myself) refer collectively to "The American Farmer," I know what follows will contain a lot of hot air. There's no such thing as *the* American farmer. There are grain farmers (and there's tons of difference between corn, rice, and wheat farmers), sugar beet farmers, dairy farmers, hog farmers, sheep farmers, vegetable farmers of many kinds, irrigation farmers, dryland farmers, organic farmers, chemical farmers, greenhouse farmers, urban farmers, market garden farmers, horse farmers, fish farmers, cattle farmers, hop and malting barley farmers, small-fruit farmers, part-time farmers, full-time farmers, make-believe farmers, pot farmers, sunshine farmers, moonshine farmers, and street-corner farmers. The extreme variation is why it is so difficult to unite them all into one organization. No one has ever been able to do that. Farmers are often in competition with each other, and though they'd never admit it out loud, when corn yields are low in Iowa, the corn farmers of Ohio can't help but be just a teeny bit gleeful because the market price of corn just might go up.

There are scores of farm organizations as a result of the diversity. Most of them, like the Ohio Ecological Food and Farm Association (OEFFA), are relatively new, very activist, and growing in membership. The older Farm Bureau, National Farmers Organization (NFO), and National Farmers Union are three groups that try to gain members from all walks of farming life. Farm Bureau has traditionally been the organization of choice for the fatter cats, while the Farmers Union and NFO were safe harbors for those struggling to make their next land payment. From Farm Bureau I once got a letter scolding me for what I had written. From Farmers Union, I got an award. To show how times are changing, one local chapter of the Farm Bureau recently invited me to give a speech. Talk about your amazing grace.

Farm Bureau is by all accounts the largest general farm organization, but, ironically, their very success may backfire one of these days because the group's policies have for years favored "get big or get out." Now the big farms are starting to swallow up one another, which means fewer Farm Bureau members. And the biggest corporate farms, like the little artisanal farms, don't have much interest in joining any traditional farm organization. So the Farm Bureau has wisely started a new public relations drive, which piously proclaims that it is for all farmers, even the (ugh) small, part-time, organic ones it has snubbed for years. Ho ho ho, we are suddenly just one big happy family. Although the hypocrisy involved here sort of gags my guts, I actually think it is a good idea—another sign that the face of farming is changing. I know some astute Farm Bureau members, and maybe they will have the good sense to push artisanal garden farming as enthusiastically as they support that agricultural nightmare, corn-based ethanol.

Even within the same category, farmers vary all over the place in terms of personality, lifestyle, and political philosophy. They are generally like most other human beings and, unlike the way farm magazines often portray them—standing staunch and stalwart out in a field like the Archangel Gabriel on Judgment Day—they love to sit around in cafes lying about their corn yields. They can play a mean round of golf, go crazy at football games, read books, sit on hospital boards, or keep peace in the neighborhood and an eye out for thieves, almost as effectively as sheriff's deputies can.

Most large-scale commercial farmers tend toward conservatism and vote Republican, but don't take that for granted. One of the biggest farmers I know (about eight thousand acres of corn and soybeans and counting) is a staunch Democrat. He says he always makes more money when a Democrat is president. On the other hand, an organic farmer friend, who is very critical of agribusinesses like Monsanto and chemical farmers in general, is a confirmed Republican. I like to compare both of these guys in my mind to another farmer in these parts who with his wife makes a living by garden farming about two acres and who leans philosophically toward Buddhism.

The most image-busting farm I know about (as I write in 2016) is Will Witherspoon's Shire Gate Farm, a 660-acre spread raising Animal Welfare Approved (AWA) cattle and chickens on pastures, with no added hormones, sub-therapeutic antibiotics, or feed additives. Witherspoon is a retired pro football player using his money from that part of his life to show the public there should be a vital connection between sports well played and food well grown. He wants to demonstrate that healthy food from healthy animals can be produced very cost-effectively without polluting the Earth. As of this writing, he was selling packages of filets and strip steaks *on Amazon* (my emphasis) at 25 percent off list price and had a sale of his AWA ground beef at $2.16 a pound at the Super Bowl in 2015.

Most industrial grain farmers truly believe—and really, really are convinced beyond all argument—that they are doing it right with all those acres of corn, wheat, and soybeans: neat and lush and weedless (sometimes), the soil worked to perfect seedbeds, the planter so refined it can insert exactly 33,450 kernels of corn into the soil per acre. Genetically modified (GMO) crops are just fine in their opinion. On the other hand, the little micro-farmer with his ten acres of organic, artisanal crops is just as firmly entrenched in the opposite opinion. GMOs are satanic. Those big stretches of perfectly worked soil are an invitation for eroding gullies. I talk to both sides nearly every day and the difference between them, I fear, is too profound to ever change completely. Maybe farming is a religion. Everyone knows he or she is right.

There are enough younger people interested in getting actively into garden farming and growing artisanal foods that by the time I figure out how to categorize them, they won't be young anymore. In fact many of them aren't young now. These newest farmers are quite the social phenomenon. Even the government has discovered them. The USDA grandly announced a new website in October of 2015, the purpose of which is to instruct interested young people in how to get started in farming. Agricultural experts will no doubt repeat the commonplace advice that beginning farmers have been hearing for a century or more. The advisors mean well. It is just that trying to tell someone how to succeed in farming is about like trying to tell someone how to succeed as an artist, writer, or musician. There are many ways, but the rules all start with the word "if."

In my experience, the person who is going to succeed in farming (or art) is too independent to go around asking for help from either educators or scientists. He or she just starts doing it. He or she has a calling, a vocation, I believe, not shared by the majority of humankind. If you read too much about what you have to do to succeed, you will be too discouraged to start. There are too many "if"s involved. But a good sign of an up-and-coming young farmer is whether he or she likes gardening. Gardening? What does pushing petunias have to do with saving the world from starvation? Well, if a prospective farmer has had the opportunity and wherewithal to garden and hasn't done it, he or she is not likely to succeed as a commercial farmer. Prospective garden farmers should be strongly urged to live with the hoe for a couple of years, during which time they will learn many of the rudiments they need to know much better than they would learn from books of instruction. Coming to grips with a hoe will teach them all about coming to grips with the unpredictable life of facing off against nature. If they have tried gardening but given it up, I doubt that the farming life is for them.

To succeed, it first of all helps to be stubborn to a fault. Bullheadedness is the common denominator of successful farmers. After that, the ones who do succeed are almost always quite suspicious, even prejudiced, against formalized education because educators have historically been so prejudicial against them. A fairly new book, *The Shepherd's Life* by James Rebanks, about farming and shepherding in the mountains of northern England, says it well:

> *I realized we [farmers] were different, really different, on a rainy morning in 1987. . . . I was thirteen or so years old. Sitting [in a classroom] surrounded by a mass of other academic non-achievers listening to an old battle-weary*

teacher lecturing us how we should aim to be more than just farmer-workers, joiners, brickies [bricklayers], electricians, and hairdressers. . . . Her words flowed past us without registering, a sermon she'd delivered many times before. It was a waste of time and she knew it. We were firmly set, like our fathers and grandfathers, mothers and grandmothers before us, on being what we were, and had always been. . . .

I argued with our dumbfounded headmaster that school was really a prison and 'an infringement of my human rights.' He looked at me strangely, and said, 'But what would you do at home?' Like this was an impossible question to answer. 'I'd work on the farm,' I answered, equally amazed that he couldn't see how simple this was. He shrugged his shoulders hopelessly, told me to stop being ridiculous and go away. . . . I thought about putting a brick through his window, but didn't dare. . . .

The idea that we, our fathers, and mothers might be proud, hardworking, and intelligent people doing something worthwhile or even admirable was beyond [our teachers].

Mr. Rebanks quit school at age sixteen and contrarily started reading every book he could get his hands on when not working on his father's farm. He is still farming, but he managed in the meantime to graduate from Oxford University, proving, if nothing else, that farmers are just as intelligent as anyone else.

I had a similar experience. Studying philosophy and theology in a seminary, I also worked on the seminary farm and realized that I liked farming a whole lot more than belaboring the heady religious theories of Thomas Aquinas. But because I got good grades (which was why I was mistakenly influenced into going to seminary in the first place), I was marked for advancement. My professors wanted me to go to Rome to complete my studies. All I wanted to do was keep on working on the seminary farm. I remember so clearly one day when I was building a fence along a new pasture we were opening up for our cows. One of the professors came out to help me. I was impressed, but he had other intentions. He was trying to persuade me to go to Rome. When I said that I was thinking about going back home to farm, he looked at me pityingly and told me I was throwing my life away, that I was wasting my talents. I went home anyway—the most talented decision I ever made.

Don't hope for much help from big business or government if you are venturing into farming because neither is quite ready to admit that bringing young people into farming is going to require shaking up the aristocracy of wealth that owns much of the farmland now. Rich investors who have put their money in farmland know it is a good risk against the uncertainties of today's economy, and they aren't likely to give it up willingly. That means your biggest obstacle is finding land you can afford. To get some of it, be prepared to practice the contrariness and obstinacy of a Corriedale ram. Land is such a precious thing that China is making more of it, dredging up the ocean floor to make new islands. Some contrary garden farmers are doing something similar, literally making their own soil out of composted wastes and farming it in hoophouses.

Government and wealthy benefactors can help finance new ideas in farming, which is good, but neither entity offers the best help for you as an individual. First of all, you will have to waste so much time going to meetings and filling out forms and listening to instructors who don't always know any more than you do, that you can't get your work done. By their very nature, wealth and power usually deal with social problems as if one size fits all, which is emphatically not the case in agriculture. Young would-be farmers differ in what they expect in the way of farm profits, some wanting to farm mostly for the pleasure of it and others dreaming about owning the whole state of Iowa someday. Their personal interests and inclinations vary all over the place, too. One may see his farming life as a milker of cows, whereas another wonders if there might be money in squeezing the juice out of milkweed (and don't laugh too hard because milkweed juice does have medicinal and technological possibilities).

People entering agriculture also differ immensely in terms of cultural background. Way out at one end of the spectrum, there is a growing number who are actually artists of one kind or another, or at least people with an artistic bent towards music, literature, sculpture, or pictorial illustration. Think of the famous Impressionist painters of the early twentieth century, who were very serious about their gardens and immortalized them in paint. Some artists, musicians, and writers today are seeing ways to combine farming and artwork, since farming, like gardening, is very definitely an art as well as a science.

If I wrote that there was a new opera being performed that was inspired by "Old MacDonald Had a Farm," you would justifiably think I was making fun. Yet something that seems even more unlikely happened in 2014. An opera based on and inspired by the lives of two of our most beloved agrarian homesteaders, Harlan and Anna Hubbard, made its debut. It was written by a young professional musician, Shawn Jaeger, and performed by trained singers and actors. Jaeger, at the time, was working at Hearty Roots Community Farm near Red Hook, New York, growing food for subscription-paying customers, a job he says he thoroughly enjoyed. "I like the idea of being rooted in a specific place, and sometimes it seems to me that contemporary music possesses a kind of generic or anonymous international style that ignores local traditions."

Shawn was approached by Dawn Upshaw, the artistic director of Bard College Conservatory of Music, located not far from the farm, to write a piece of music for her to sing with the Saint Paul Chamber Orchestra, and secondly to write an opera for her students to perform at Bard. (Bard, by the way, offers courses in small-scale garden farming.) He had met her at a professional training workshop at Carnegie Hall, which pairs young composers and young singers to collaborate on new vocal works. "It was a huge opportunity for me," recalls Shawn.

The song he wrote for her was mostly inspired by Wendell Berry, surely the father of the new, small farming trend and the premier example of someone who has combined farming with the art of writing. Shawn called his song "The Cold Pane." His opera uses sets from five of Berry's poems, including the one by that name. He titled his opera *Payne Hollow*, the name of the Hubbards' homestead farm. "Mr. Berry was very gracious and helpful. I wanted to say something about society today, about climate change, about sustainability and what I consider to be the way forward for our species, without being heavy-handed," he says. "I thought the Hubbards represented a powerful, peaceful example of doing just that in their personal lives." While the Hubbards, now passed away, lived primitively in terms of money and derived most of their food from their own garden farming or from trading fish they caught in the Ohio River for other food, they lived in refinement on the intellectual level. Harlan played the violin and Anna the piano, and if you were a mouse in their attic you might hear them in the evening discussing some esoteric philosophical theory . . . in French. What little money they made came from Harlan's paintings and a couple of books he wrote, plus a little rent he received from a house he had built in his earlier years. I asked him once why he didn't write a book on how to live the way he and Anna did. His reply: "It would require very detailed instruction, too tedious to interest any publisher."

Jaeger's opera is only one of many honorable bows to the work of Wendell Berry. For example there's a new documentary out, titled *Look and See*, that focuses on food and agriculture specifically from Berry's traditional rural perspective. The Los Angeles director, Laura Dunn, worked with musician Kerry Muzzey to come up with music to fit the intimate kind of farm stories that Berry is famous for.

To go to the far opposite end of the new wave of farmers, I need look no farther than the two young families in my neighborhood who make hay on my little fields because I am no longer physically able to do so. They are totally farmers of the nuts-and-bolts variety who have never listened to an opera and probably wouldn't want to. Their backgrounds and farming activities are different, yet their lifestyles are quite alike. Matt and Nelson are brothers-in-law. Matt and his father and grandfather farm together on a large, successful commercial grain farm. I remember when Matt's great-grandfather's barnyard was full of mules. He was taking them in trade for the Oliver tractors he was selling as he shifted from milking cows to a farm machinery business. One son carried on that work; the other stayed in farming. In his spare wintertime, right now as I write, Matt works part-time in an auto repair shop and keeps a few chickens and livestock on his home place. Nelson is more of a garden farmer with a few acres populated with chickens, sheep, a cow or two, a horse or two, and quite a herd of used machinery, which he keeps accumulating in hopes that someday he'll be able to buy more land. He works in town for a tire service business and has already bought another ten acres.

Although their farm situations are economically far apart, Matt and Nelson's home lives are similar. Both are busy with their wives raising families. Both enjoy keeping alive the old ways in farming by raising a few chickens and livestock and using old farm machinery of my era on the "garden" part of their farming. Eerily, Matt views his little home place operation as an activity not really part of the large-scale grain farming from which he makes a living. They would not even be able to get back to my hayfield with today's big machinery unless we cut down a considerable number of trees, but they can snake their way through with the old equipment. Some days when I go back to the field to cheer them on, both families—wives and children, too—are busily at work baling hay with ancient equipment. I feel like I have stepped back into my childhood in the 1940s. It is just so idyllic and gratifying.

Just within my own extended family, we are blessed with about every kind of young and upstart farmer you can think of. My eye doctor, Kip, and his wife, Mary, who is my niece, are enthusiastic farmers in their spare time. I am amazed at how much he knows about farming for someone not raised on a farm. He has made it his business to learn everything he can about raising corn and soybeans—he knew what corn was selling for on the Chicago Board of Trade the last time he was peering into my eyes looking for cataracts. Seeing the economic weaknesses in industrial grain farming, his latest project is experimenting with heritage hogs. Sometimes I get to laughing as he commiserates. Here is this young and successful optometrist complaining and moaning about farm profits or the lack thereof as authentically as any crusty, old sodbuster. His wife, Mary, is good at that, too. She is thinking about raising quail for meat. Their children are taking courses in agriculture. One has graduated from college with a degree in agriculture and is angling for a job as an intern on a farm that sells its production directly to customers and restaurants.

It is unseemly to brag about one's own children, but it would be even more unseemly to ignore our son and daughter-in-law, Jerry and Jill. Jerry is in the construction business, and Jill works for a local accounting firm in town. (One of the partners in that business, John Mizick, has been a garden farmer on the side since before that phrase came into use.) Jerry and Jill are two of the hardest-working people I know. Neither went to college. Jerry started doing hard, physical construction work—roofing buildings, for example—when he was barely out of high school, and Jill pumped gas at her grandfather's service station when still in high school. They eventually had a chance to buy about forty acres reasonably priced because it was mostly not very good land and belonged to family. Jerry built a beautiful brick house on it, and although his spare time from construction work is very limited, he and Jill built two barns, raised sheep for a while, and now pasture steers to produce grass-fed beef for family and a few other clients. The rest of the farmland he rents out to neighboring farmers. He heats his house almost entirely with wood he cuts himself.

A nephew, Johnny, always wanted to be a farmer, his mother tells me, but buying one was out of the question. He worked at various jobs and finally landed a good position in electrical engineering. Again, no college. He saved his money. An opportunity arose. A retiring farmer was starting to talk about selling some of his land. It is difficult to buy smaller tracts of farmland in our county—the competition is keen between the larger farmers. Johnny was right there, checkbook in hand. After lots of hemming and hawing, the owner realized Johnny wasn't going to go home until he had an answer. Sometimes you've got to get very bold and bullheaded. Root, hog, or die. Scratch, hen, and live.

Most of the forty acres Johnny bought was in rather rough pasture and brush land. Over the next five years, an astounding transformation occurred. He plowed up the pastures and replanted them to productive grasses and clovers. He stretched new fences, built a bridge across the creek, dug a beautiful pond, built two barns and a house—he and his wife Julie lived in a camper and used an outhouse until the house was ready to move into. Even though they both work off the farm, they raise sheep plus a few chickens. Johnny has built up quite an inventory of used equipment, much of which should be in an agricultural museum. But he is adept at keeping it going. He ran an irrigation pipe to his big garden and truck patch, something unheard of around here. He keeps at it day and night, and some of us fear for his health. I tell him how I marvel at his energy and often ask him if he ever gets tired. His answer is always, "I just love it." He enjoys tinkering with old machinery, a characteristic that I suspect is essential to the success of small farming operations. He and Julie sell eggs, pumpkins, lambs, wool, plus corn and straw when he has extra. But making money is not his goal. He is enjoying life.

Another of my nieces, Abby, is married to Tom Smith, who used to be head chef at The Worthington Inn on the outskirts of Columbus, Ohio, one of the area's most celebrated restaurants. He was one of the early pioneers in buying food to prepare in the restaurant directly from garden farms. After he and Abby moved to their own little garden farm, he took their produce to the restaurant to use in the gourmet salads he was well known for. His latest venture is a pizza parlor that he has opened in our village. He is interested in bringing gourmet artisanal food to the beer and hot dog crowd.

Jandy's farm market, operated by our friends Jan and Andy in another part of the state, is an example of the kind of "commercial" garden farming that makes just enough profit to maintain stability without the hassle of expansion in today's paper money economy. I put quotes around commercial because, although Jan and Andy make their living selling what they grow, "commercial" is about the last adjective I'd use to describe them. We have watched them make a living off hardly more than two acres for many years now. Most people, including many farmers, do not think that is possible. Andy did have a job and built up a nest egg before he bought their land, and they have lately inherited a little more, but they have lived off their garden farm earnings quite comfortably all these years. As they point out, it is easier for them because they have no children to raise. The two secrets of their success are that they are consummate organic growers and they rarely spend even a penny ostentatiously or unnecessarily. Their house is humble and built partially underground to save energy, their clothes are plain and simple, their food comes almost entirely from their gardens (they are vegetarians), and their home fuel comes entirely from wood they cut and split from their own woodland. Their main commercial crops are onions, garlic, lettuce, and cut flowers, which Jan is adept at arranging into attractive bouquets—combining art and garden farming literally. But they raise other food crops, too, so that they have something to sell every week at their local farmers' market and at their own annual Garlic Festival. Their produce is now so well known locally and so high in quality that some Saturdays they are practically sold out well before noon.

Although all farmers are different as individuals, there are several economic categories most of them fit into. First of all, there are the ones who have managed to survive by "getting big." Anyone who rose up through the farmer ranks over the last fifty years without inherited money, to own and operate several thousand acres or more of corn and soybeans or cotton, has to be by definition about as smart as Einstein. In fact, almost all of them did start out with some inherited land. It would be naive to criticize them because of their preference for big acreage, big machinery, and GMO products. Without these high rollers, the food business would, right now, surely be in more trouble than it is. But these farmers had to join The Economy to accomplish what they did, and so became part of the larger money society. Several complain to me periodically that their success has ensnared them in the money trap. "We are becoming little more than foremen for huge agribusiness companies, not independent farmers," one of them said to me recently. "We are so far into Monsanto-type farming that we can't get out even if we wanted to." They are becoming, in fact, wealthy serfs.

Some in this group are still contrary enough to maintain some independence by getting into bio-intensive organic methods. They are very daring risk-takers, still believing in the "get big or get out" philosophy. Seeing money to be made in organic meat and milk, some of them build big animal factories, hoping to produce large volumes of high-priced organic foods. Their business plan is to buy their organic grain and hay elsewhere, which would solve the almost impossible task of trying to grow crops organically on a large scale. Lately they have been in trouble financially because weather and demand have cut the supply of organic feed and they have to pay high prices to get it. At the OEFFA Conference 2016 in spring, John Bobbe, executive director of the Organic Farmer's Agency for Relationship Marketing, told me that 40 percent of the organic grain being fed in the United States is imported. Let that statement sink in for a second. That means paying a prohibitive nine dollars a bushel or so for corn right now, when the price for non-organic corn is under four dollars. This is just more evidence to me that organic farming must remain small-scale. Smaller garden farms can produce their own feed with their own labor for their comparatively small number of animals and so avoid this dilemma. It is also a surer guarantee to the consumer that the milk and meat really are organic.

The rest of the farmers fall into four general economic categories, more or less (notice how I hedge once again). First are the ones who have always farmed conservatively in terms of money and have expanded very slowly or not at all from a few inherited acres, rejecting borrowed money almost always. The secret here is to have a son or daughter willing to take over and still live a financially conservative lifestyle. These farms may be organic or transitioning to organic or they may not. They keep expanding in production a little, not in acreage or number of livestock so much as in quality: higher yields from the same acreage, higher milk production per cow, or specializing in pedigreed or purebred breeding stock or some special·crop that sells above normal market prices. This kind of farming can't be jumped into overnight but must be built up slowly over more than one generation, inspired by a long and disciplined love of the land. Many of these farmers are quite happy to live unostentatiously and save every penny they can. Often by old age they are embarrassed to find that they are millionaires because of the value of their land. In years where there is little profit, they can still weather the storm with help from earnings their savings accrue. They farm because they love the lifestyle, not to make big money and move to Florida.

A second group is composed of farmers who have another job or business that fits in well with their commercial farming. The goal of some in this category is to make the farming enterprise profitable enough so they can quit their other job. Sometimes they are successful, often not. They work so hard that I fear, and tell them so, that they are endangering their health. But many of them actually enjoy their work off the farm as well as on it.

A third category is made up of people who garden-farm mostly for the pure pleasure of it while doing something else to make enough money to live on. Some of them dream about their farm work making enough money so they can quit their off-farm jobs, but they don't do it primarily for that reason.

A fourth category, somewhat like the third category, is spearheading the new wave of artisanal garden farmers. They give every indication that they might become important specialty food producers in the future. They are very excited by all the possibilities that the local, decentralized food movement is bringing to the marketplace and especially by the new crops or foods that are sparking this market. They aim to make money at farming, but intend to keep on doing it, even if they don't.

Some home sites show telltale signs of who lives there. Like the warning "Beware of Dog," a rick of split wood near the house says "Beware of Contrary Garden Farmer." Scott Nearing, surely one of the fathers of not-necessarily-for-profit garden farming (although he made quite a profitable business out of maple sugaring) cut his own wood until age one hundred, when he figured he had enough laid by to last him out. There is always a rick of split wood close to Wendell and Tanya Berry's back door, and I notice in his picture on the front of *Humanities* magazine (May/June, 2012) that he is standing in front of his woodshed full of wood.

Running through all the categories there is an impulse or motivation that is not necessarily good, I fear. I know it is true of myself. I see my little farm as my island of security when in fact security does not exist anywhere in the world. I am in danger of deluding myself into believing I can live apart from the turmoil and be safe. Monks try to do that, too, and learn soon enough that worldly insecurities burrow themselves right into monastery havens. I know because I tried that, too.

CHAPTER 2

Farming Is All About Money, Even When It Isn't

The biggest mistake in getting into farming ventures is the assumption that, if you know how to do it and follow the know-how strictly, you will succeed. A steady flow of how-to books, college courses, and online instruction supports this fancy. This information is necessary, of course, but it is just not the first requirement. You can learn all the scientific and technological facts of the matter and hone them to perfection as you work, but the weather can always make you look like a loser. And even if you find a way to make a fortune, you will not necessarily be truly a successful farmer. In fact, there are very "successful" millionaire farmers whose personal lives are a shambles.

After you acquire the habit of eternal contrariness, you must next try to master the philosophical how-to of following a way of life that does not respond well to the usual industrial workplace formulae for success. You must believe that happy farming is about putting love of the land before love of getting rich. Actually, the philosophical how-to is almost a Ten Commandments sort of thing, or a book of ethics. Or a summation of all the wisdom that wise people have uttered, written, sang, and preached since the dawn of history and probably before that. I list them hesitantly, knowing the reader has heard these prosaic admonitions before.

The difference is that now when you learn to do without something, you save a lot more money than in previous times because stuff you don't need costs a lot more money.

1. Unless you are rich or inherited your land, you almost always must have another source of income to succeed or even just to survive on a garden farm. Even in the heyday of the agrarian society of the late 1800s, almost all farmers did more than farm. They were schoolteachers, doctors, lawyers, merchants, plus doing skilled work on the farm that we now pay others to do. So, too, with many farmers today. Even some of the Amish have off-farm jobs to help the cash flow. It is the economic nature of the beast. Real-money profits from farming just come in too slowly to jive with our industrial, money-interest economy. The modern notion that it is beneath a real farmer's dignity to have another job is something left over from the landed aristocracy of the Middle Ages. Almost all farmers today have more than one source of income—even, or most of all, the large acreage farmers. One of my cousins started out penniless and on his way to becoming a very successful large-scale farmer; he operated a blacksmith shop for awhile, drove trucks for awhile, farmed with his father always, eventually owned a restaurant, a motel, and a stone quarry, using profits from these businesses to increase his farm size to something like eight thousand acres today. When I chide him for continuing to buy and farm more land, he answers: "I will quit farming more land when you quit writing more books."

2. Don't go overboard on anything. Oh, maybe once in awhile, but when you jump in the water, keep a firm hold on the boat with one hand so you can get back in when the fervor passes. The rule, as wise people have written forever, is moderation in all things. Two acres might be enough. But don't get too fervent about it and don't criticize big farmers too much. There will always be bigger farms, even huge farms, to answer needs and markets small farmers can't fill or to try to satisfy human ambition. They at least perform an ethical service by sometimes going under and making the land available again to young buyers. The meek shall inherit the Earth.

3. "Neither a borrower nor a lender be." Shakespeare said it first. And also remember the lines that follow that one: "For loan oft loses both itself and friend/And borrowing dulls the edge of husbandry." When you must borrow, like for your first house, refer back to No. 2.

4. Before you start smoking or drinking, do a little math. The money you save in a lifetime from avoiding these habits (and don't forget to add in the interest earned by the saved money) is awesome. If you like a beer or two every day, or a shot of whiskey or two like I do, soon you will be edging your way up to five or six beers or shots every day. Do the math over a lifetime and see how the money saved from drinking only moderately or not at all can mean a nice little chunk of money, not counting what you save by being healthier. Other kinds of small savings can add up, too. My wife, Carol, has cut my hair for fifty-five years now. If you get ten haircuts a year for fifty years at an average of $5 each, that's $2,500 saved plus interest. Piffle, you might say. But if that $2,500 is money you did not have to borrow to help pay for a house loan on which you would be paying 4 percent interest, let us say, for thirty years, that's $3,000 more saved, if I've got the math right.

5. Don't buy anything you don't really need. My mother constantly preached to us the old saying: "Use it up, wear it out, make do, or do without." The saved money can buy time for work or play you really enjoy.

6. Avoid long-distance travel as much as you can. It costs lots of money, and we are approaching a time when we will all be able to "travel" anywhere in the world (and beyond) inside computer headsets. One of the money-making advantages of husbandry is that it limits the amount of traveling you can do. You have animals to take care of. And they seem to know when you are away and that's when they get out. Staying home is not much of a sacrifice for garden farmers anyway. One of the signs of a true farmer is a preference for staying home.

7. Eat mostly at home. That's easy enough for me to say since my wife does the cooking. But you can save a heap of money by

just not eating breakfast and lunch in a restaurant. Stay home and eat corn flakes. I love corn flakes, especially with my sister's homemade granola sprinkled over them. Her granola has hickory nuts in it. Then I add honey from the apiary down the road and flood the whole bowl with cream. Yum. Our cream used to be from our own Guernsey cow. Super yum.

8. Cultivate a good mechanic and be content with driving an older car. A new $50,000 car loses about $5,000 the first time you drive it around the block, as we all know. If you habitually drive cars on borrowed money you are losing your hard-earned cash faster than if you were gambling in a casino. Buy a used car and pay cash saved from not borrowing for a new one. My pickup I bought new in 1981 with saved cash, and it still runs well because I have driven it only about forty-five thousand miles. If you are paying out even 5 percent on $20,000 car loans for sixty years, that's like $60,000 just in interest; and, again, don't forget to add in the interest that would accrue if that money were in savings. Okay, so right now there's very little interest on savings. This is how The Economy is punishing savers whom it otherwise hypocritically encourages. Savers aren't doing their part to keep the spendthrift economy going. If you think I am only joking, look at what banks in some countries have been doing—charging savers a fee for depositing money. And the bankers sometimes come right out and say it: Savers don't contribute enough to the paper money economy, so they have to be forced to contribute. In retaliation, savers buy stocks and bonds to make their money "grow," and if they lose it that way, serves them right for saving instead of spending.

9. Life's biggest expense for the lower and middle classes is paying back money borrowed to buy a house. Some homeowners pay interest on home loans most of their lives and the total amount is ghastly. Getting out of big debt as soon as you can, even at great sacrifice, is the easiest money you can ever make. Yes, home loans are deductible from income taxes, and it might make sense for rich people to stay in debt when they wouldn't have to, but not the rest of

us. Do the arithmetic. Even at low interest rates of 3 to 4 percent, a $200,000 home eats into your earnings by around a hundred dollars a week. Be satisfied with a smaller house. Use the money saved to buy your garden farm land. Sometimes you can buy a modest house with several acres for less money than a big, fancy house with no land. Remember that you can always make your house fancier after you move in, using saved money, not borrowed money. Remember, too, that the bigger the house, the bigger the utility bills.

10. One of the most lucrative aspects of garden farming is that you don't need a college degree to do it. I will catch hell for saying this, but many young people are going into steep debt to get a degree they don't need. If you read voraciously all your life, that is about as good an education as four years spent frenetically in a college. Learn how to farm by doing it. No telling how much money the Amish save by shunning college, money that goes round and round in the community to juice up the local economy. And my Amish acquaintances are as well informed as most college graduates I know, including myself—and can buy and sell most of us, too.

11. Be content with cheaper clothing in general. Sometimes fashion actually rewards you for doing so, like the current fad for wearing jeans with big holes torn in them. But while doing farm work, don't wear cheap shoes. The bad feet you might end up with will cost you more than good shoes.

12. Do not enter into financial relationships with foolish or stupid people, even lovable ones. They may mean well, but they will hurt you every time.

13. Bargains are temptations to buy stuff you don't need.

14. When you must buy, choose quality. If the purchase is for something you truly need, you will save money in the long run. Maybe even in the short run. But everyone knows that. I think.

15. Don't necessarily do today what you can put off until tomorrow. Might turn out you don't have to do it at all.

The reader is expecting me now to say how much slow-lane money savers are opposed to our spendthrift society. *Not so.* We understand basic capitalism, if understanding it is possible. It can't function without ever increasing consumption. So be it. Most people do not want to live the tranquil, slow-money, home-based life that garden farmers are trying to follow, and that's fine. If everyone loved the solitude of the woods like I do, there would be no such solitude and we would probably have run out of trees by now. The profit that most Amish farmers achieve comes from farming in the deflated money world of the past while selling in the inflationary world of the present. Buy low and sell high. For garden farmers to live the way they do, lots of other people have to spend lots of money. Working hard to make and spend all that money makes for a difficult lifestyle, but someone has to do it.

I often think that the main reason we garden farmers live the way we do is not primarily to grow good food but because life is more tranquil not participating, or participating only a little, in the anxious money economy. We adherents of staying small and staying in are inherently lazy in our dislike for the kind of money hassle required to get big or get out. It is not possible to avoid the money economy, of course, not even for monks who have taken a vow of poverty (and, as I have said, I ought to know because I tried that, too). The only way to escape the jaws of money is to die. Until then, the trick is to stay out of the place where the jaws come together. I quit giving speeches and avoided long-range travel because these activities destroyed my peace of mind. But that meant giving up a considerable amount of income. I quit a good-salaried job for the same reason and that lost me more. But I did not feel sacrifice, only freedom to stay home and enjoy life. I didn't like being afraid of getting fired for what I might say. I didn't like forever watching the clock to tell me when I could go home to my real work. I didn't like working for someone else or having other people work for me. Achieving the so-called American Dream made me unhappy.

Garden farmers have other reasons for at least trying to reject The Economy. First of all, there is a kind of logic of the inevitable involved. If you go whole hog for money, you risk losing whole hog, too. Even if you inherit a fortune, you have to be careful. The Economy is going to fall on its face about every twenty years, come hell or high interest rates. Economists call these "events panics" or "recessions" or "market corrections." How about "periods of temporary lunacy?"

But money is so handy, especially if you have a lot of it. So borrow our society will, even when it's not really necessary. The ultimate result is that our collective embodiment, our nation, must borrow more money than it takes in, too. Even though common sense insists that no one can get away with that forever, the United States has been doing it from its very beginning. Perhaps there is some kind of backward cleverness involved. Perhaps we must live beyond our means as a way to maintain our means. A few brave economists are suggesting that the government should just do away with debt ceilings. All we have to do is pretend that money really does exist outside the mind, and we are saved. Some 80 percent of the world's population seems to be happy believing that the gods that exist in their minds also exist immortally outside their minds, so why not immortalize money the same way? Yes, I'm kind of joking, or at least I thought I was until I came across a recent book by a respected economist, Adair Turner, titled *Between Debt and the Devil.* Turner advocates very seriously something like what I jokingly suggest here.

Garden farmers hold on to the dream that there might be a way to live contentedly with only a little of this pretend money since no one seems to live contentedly with a lot of the stuff. The way they view "making a living" is not really much different than what small-town store owners do. I know any number of shops operated by people who just like to "play store." If they break even, they are satisfied. Or sometimes one of their spouses has another very lucrative business or fat bank account, and they want a way to lose a little money for tax purposes. But most of us have to make at least some. The first question young people pining to live like I do always ask: Is there a way to have a farm without being rich?

Yes. The first step is to find someone to live with who agrees with your contrary philosophy of "root, hog, or die" or "scratch, hen, and live." It is almost always better that at least two people work together on even the smallest farm because one of them at least will have to be employed part-time off the farm. But finding a partner can be difficult. Imagine a young man courting his beloved today, saying:

> *I don't care much about money. In fact, I hate the stuff. I plan never to go in debt for anything except maybe a piece of land and a modest house, nothing fancy, just low-cost and energy-efficient, maybe even build it myself, something I can pay off in a few years and then be free of debt. I want to provide my own food, fuel, fiber, and recreation, not try to make more money than I absolutely need—cheap car, cheap clothes, cheap pastimes, cheap vacation. Or none at all because my whole life on my homestead will be my vacation. Will you marry me?*

Obviously, this is not the way to go about it, although if you find a spouse crazy enough to say yes to that kind of proposal, you are well on your way out of tinsel money slavery. Otherwise, with a little luck, parsimonious ingenuity can eventually win the day for you. Young people tell me that they can't afford to buy a little piece of land like Carol and I did. Today's land prices are too high, they say. But land prices are always too high. We bought our twenty acres in 1974 for $700 an acre and thought that was too high. Later we bought ten more for $2,000 an acre, and I know that was too high. When pioneers were paying $5 an acre for homesteads, they thought that was too high. Money is relative. You have to think outside the tinsel and adjust your dreams to reality. If the price you have to pay for twenty or thirty acres really is more than you can afford, consider a five-acre homestead. Five acres can be more than enough to keep you busy and independent if garden-farmed to the hilt.

When I try to define the strange attitude that garden farmer types assume when they seem to shy away from the money-striving way of life, I think of the famous economist, John Maynard Keynes, of all people. If you have ever tried to read Keynes closely, you know that he manages to say everything that can be said about money, up one side and down the other. On any given page he may seem to be promoting one idea, and on the next page an opposing idea. It is difficult to put him in anybody's philosophical camp. His most mystifying idea in this regard he turned into an essay in 1928, "Economic Possibilities for Our Grandchildren," which still gets him in a lot of trouble. In it, he mused that in a hundred years, technology would render human labor for the necessities of life all but obsolete. The big problem would be what society would do with all that leisure time. Viewing what has happened, Keynes' musings seem ridiculous, of course, but when you read him through, he is actually thinking like a garden farmer. He assumed that people would learn to take advantage of the new freedom from everlasting wage work that technology would provide. They would relax, save money, and smell the roses (or the barn manure). When he learned that people preferred to spend rather than save, he introduced the idea of compulsory savings into economics, sort of a forerunner to our Social Security. In 1941, he suggested that the government (in England) should take a part of every man's income and put it in National Savings Certificates. People would learn to enjoy "the art of life itself," he wrote. He did not seem to grasp fully the acquisitive nature of the human mind and how the majority of people would always find more things that they thought they must have, and so they would always be willing to keep on working and borrowing madly for the money to buy them. They didn't want leisure time; they wanted to keep on buying. Garden farmer types understand Keynes, or would if they read books on economics. They want leisure time not needed for money-making so they can stay busy doing things that make for art and satisfaction without money.

Is there a way to get there? Read on.

CHAPTER 3

_∼_

The Economic
Decentralization of
Nearly Everything

G arden farmer types shy away from crowds as assiduously as
they avoid poison ivy. A crowd is just a heartbeat or two
away from a mob as we are seeing all the time now. The
more people crowd together, the more irresponsible and irrational
they tend to act. The more they stand out courageously as individuals,
the more they try to, or at least pretend to, exhibit concern about the
results of their actions. As individuals, they can't hide in the crowd
the way a stockholder can hide in corporate irresponsibility. Concern
about the groupie mentality is one reason that decentralization of
nearly everything is gaining momentum. The garden farming part of
society is shying away from the mob mode. (But of course, if everyone
shied away, shying away would become crowded, too. That is why it is
better that the new farming movement is a quiet "unrevolution," not
a shrill and loudmouthed revolution.)

Even after I first saw clearly that there was an unrevolution
in progress, I nevertheless resisted using such grandiose terms
as "looming on the horizon." Maybe it was just a fad puffing up
briefly in a whirly-gust. Or a blip on the Twitter-scape. Perhaps
all those people with enough money to pick and choose fancy

foods would tire of going to the new farmers' markets that were springing up everywhere, or to the supermarkets and restaurants that were concentrating on organic or gourmet food from local artisanal sources. I was afraid to believe what I was seeing and kept telling myself that money would always rule, and therefore, mass-produced factory food would always be the norm.

There are plenty of arguments to support that point of view. If the accountants worry their books a certain way, define terms in language that underscores industrial economic assumptions, take advantage of every subsidy and tax break that the factory food producers have available, mass-produced food can seem to be indeed cheaper and maybe even good for you. Better than starving. The champions of capitalism and socialism both can boast about how fortunate all those poor people are—factory farms can supply them with enough cheap fast food to cause them to die early from obesity.

But now, as I write in 2016, there is clearly something new and invigorating going on in the food production world as artisanal garden farms multiply and large-scale industrial farming fades into the fog of crushing cost overruns and faltering prices. The new economy understands that farming is a biological process, one to be handled with careful love and very gentle agronomy and husbandry, not industrial production that concentrates on cramming more and more animals under one roof to lower the per unit cost of production, or growing corn on hillsides and prairies that nature never meant for industrial cultivation. Significantly, *Farm and Dairy,* a publication in eastern Ohio that regularly supports conventional farming and has in the past been rather aloof toward the practicality of a garden farm society, recently ("Can't Ignore the Trends in Agriculture" by Susan Crowell, January 20, 2016 issue) cautioned its readers that they "don't want to be blind to a trend that could bring massive changes to your farm or ag business. . . . Consumers care more about how their foods are produced and sourced—and that's impacting the entire food chain. . . . Local food is no longer a fad. . . . People want 'real' food." The editorial cites a U. N. statistic

that says there are eight hundred million people worldwide practicing urban agriculture. Add to that the millions of home food
gardens in the countrysides of the world. The United States alone
has seen an increase between 2007 and 2013 of some four million
food gardens to some thirty-seven million total, and certainly more
by now. Community gardens tripled from about one million to
three million during that time. In Cuyahoga county (Ohio) at that
time, there were twenty-eight commercial farmers' markets and
way more small artisanal garden farms selling direct to consumers,
stores, and restaurants. In the Cuyahoga Valley National Park area
alone, there are eleven such farms (some using draft horses) under a
program the park called Countryside Initiative Farms.

A good way to see what is happening is to read one of the
Edible publications, now with over eighty different magazines in
its fold, covering the entire nation. This fairly new publication is
the voice of the artisanal garden farming movement. In one issue
of *Edible Ohio Valley,* I counted over seventy ads from farms, restaurants, and organizations involved in the local food business in that
area. In *Edible Cleveland,* there's a list of thirty CSA operations
(community supported agriculture) just in that area. Ohio has at
least eighteen artisan cheese businesses, sixteen of them owned
and operated by women. In the eight counties of northeastern
Ohio (I use this area because I am familiar with it—the statistics
are the same all over the more populous areas of the country), there
are, as I write, sixty farmers' markets.

Books are pouring from the presses and computers, everything from the fiery and feisty *Old Man Farming* by Lynn Miller,
publisher/editor of the *Small Farmer's Journal,* insisting that the
world can't survive without an influx of many small farmers, to
Carol Deppe's serene, Tao-inspired book of instruction on how to
grow all one's food almost completely independent of the outside
world (*The Tao of Vegetable Gardening*). It is obvious that the decentralization of the food business is not a fad. It is fast becoming a
way of life, and the very first reason is that the business model

for it, or rather the unbusiness model, does not necessarily require borrowing a lot of money to get into it. It does not necessarily require starting a business at all. The best evidence that the movement has gone beyond fad is the burgeoning number of people in urban areas who are keeping a few backyard chickens.

There are also clear-eyed business people who have looked into the entrepreneurial reverberations of developments like backyard chickens and gone into action. John Emrich, in his book, *The Local Yolk*, tells how he started a business he calls "Backyard Chicken Run," delivering supplies to backyard poultry producers in the Chicago area. Even more a sign of the unrevolution, he spends much of the book arguing that sustainable farming is one of the best money investments today—and he states this after a successful career as an investment consultant and money manager.

We are into something even bigger than it might look because of the revolution in electronic communication that the computer has brought to us—a way to compete with The Economy by not competing. That's another reason why I call it an unrevolution. Garden farming does not have to keep centralizing production into ever larger units like businesses have had to do ever since the Industrial Revolution came along. Earlier economic movements toward decentralization, like distributism, which rose in the late 1800s, were ahead of their time. The Distributists disagreed with both capitalism and socialism, arguing instead for spreading private ownership and means of production out as widely as possible among as many people as possible. For a while, their battle cry was "three acres and a cow," inspired by the government promise to give freed slaves "forty acres and a mule" after the Civil War. (Janice Holt Giles, in 1950, wrote *Forty Acres and No Mule*, in which she describes perfectly a particular kind of contrary garden farmer thriving in Appalachia.) "Three acres and a cow" was also picked up and publicized by G.K. Chesterton, and although he was hardly a garden farmer, his Distributist sentiments are still applauded by many of us who are.

But without instant electronic communication, a person could feel mighty beleaguered out there on the ramparts with three acres and a cow. Distributism as a movement withered or barely hung on. Centralization became salvation. If you wanted to be a writer, you were told to move to New York where the writing community could huddle together for survival. If you wanted to succeed at farming, you had to "get big or get out," which is a euphemism for "centralize." There are grain farms now that encompass fifty thousand to a hundred thousand acres—and larger in Russia and Africa—under one paper ownership. They live the same precarious life as that of the huge bonanza farms of the American Midwest that went broke shortly after the turn of the twentieth century (see chapter 24, "The Real Background Behind the Fading of Industrial Farming"). After you have consolidated the whole world into one giant farm and it still doesn't return its investors enough profit, what do you do next?

Decentralization has its limits, too. The way things are going in the Middle East, virtually every town in Iraq wants to be its own sovereign nation. Even in the United States, there's a popular notion that the way to solve political and religious disputes is by secession from the Union. History shows, however, that whenever any ideology vanquishes its rival, another soon rises up to replace it or the reigning ideology splits into factions that fight each other. Humans love to quarrel. The saving grace of the current garden farming unrevolution, like unrevolutions in the world of art, is that it does not depend on finagling politics or religion for its welfare. Its members know how to feed themselves, shelter themselves, and entertain themselves. They bow to no plutocracy and are in a position to endure plutocracy until it collapses on its own. Eliot Coleman, one of the pioneers of organic farming in New England, says it well in a quotation from his writing (I found it on Dave Smith's very contrary *Ukiah Blog* at ukiahcommunityblog .wordpress.com, "Why Small Organic Farming Is Indeed Radical (and Beautiful)" on February 7, 2010):

The small organic farm greatly discomforts the corporate/ industrial mind because [it] is one of the most relentlessly subversive forces on the planet. Over centuries both the communist and the capitalist systems have tried to destroy small farms because [they] are a threat to the consolidation of absolute power. . . .

It is very difficult to control people who can create products without purchasing inputs from the system, who can market their products directly thus avoiding the involvement of mercenary middlemen, . . . who can't be bullied because they can feed their own faces.

Decentralization is not quite the same thing as localization. Big businesses have local branches where they try hard to project an image of good ol' down-home folks. The current popularity of using the word *local* to describe the decentralization of food production has drawbacks. One company offers its customers "local" food from no farther away than "three hundred miles from your home." Really. And just because a food is produced even a mile from where you live does not mean that it necessarily tastes good, but only that there is a better possibility that it will taste good. If you want a good pomegranate, you might have to order one from Afghanistan. Decentralization is sort of like the father of localization. Pomegranates may not grow here now, but maybe small, backyard efforts, without having to risk big gobs of money, can figure out how to do it.

The trend toward decentralization is evident even in architecture. Perhaps you have heard of the tiny house movement or the Small House Society. There seems to be a market opening up for houses as small as 250 square feet. The savings involved in energy and building costs is enormous, of course, and apparently there are people who are satisfied living in smaller spaces—even happier in smaller spaces. It is so cozy.

Technology supported consolidation in the beginning of the Industrial Revolution, but today increasing the size of a factory does not automatically lower per unit cost in activities like producing quality food. What is required are small-scale producers who are not looking for the big profits of mass production, but only a way to provide enough money to finance their not-necessarily-for-profit lifestyles. Technology now makes it possible to stay in Deadfurrow, Tennessee, growing an ancient flint corn handed down in your family from the frontier days and selling enough of it worldwide on the internet to pay for the low costs of living a low-cost life. You may not make much money, but you won't go broke, either. All you have to hope is that United Parcel Service doesn't run out of gas.

Good examples of what is happening pop up in the news constantly. On Cape Cod in the good old days, processing and selling sea salt was a thriving backyard business. Slowly but surely, industrial capitalism pressed the small seaside processers to consolidate into larger and larger units until finally only a few companies ruled the marketplace. But drying salt out of seawater is a process that does not lend itself well to industrial processes. It requires lots of hands-on labor and love. So now the business is going the other way. New, small sea salt shops are on the increase because they can produce a quality product with less overhead than the big salt companies.

Technology is always the servant of economic profit. It dashed valiantly to the aid of centralization when that appeared to be the place that more money could be made. In the livestock world, it made possible giant installations in which five thousand sows could be housed along with their fifty thousand offspring—and never lose more than one thousand pigs a year. Eventually this kind of "efficiency" reached its limits. So now technology is switching its loyalty to the new mistress, decentralization, and coming up with all sorts of clever ways for it to shine. For example, low-cost plastic- and fabric-roofed structures of clear flexible material where pigs can be raised on sun-drenched, composting bedding are suddenly blossoming in the fields where it seems only yesterday the horizon showed a landscape of abandoned, cement-block animal factory buildings.

There is precedent for all kinds of "enclosure agriculture" on garden farms. Those walled gardens that look so quaint in England began as necessities to protect growing food from wild animals and wild humans, not to mention maintaining a slightly milder microclimate inside the walls—the stone or brick soaks up heat during the day to fend off frost at night. If you find that hard to believe, lay a brick that has a hole through it on your lawn in February. By the end of March grass will be growing up through that hole three inches tall while the grass away from the brick is barely greening up.

Then civilization went to glass houses to keep predators at bay and to extend the growing season even more. But glass greenhouses were too expensive for much large-scale exploitation. Now smaller farms can afford the plastic-covered hoophouses. New designs and new applications arrive almost daily to make it easier and cheaper to cover and uncover the growing plants as the weather dictates. A click or three on Google will bring you more information in a minute than you can digest in a week on the latest designs and improvements in high tunnel and low tunnel hoophouse technology.

What this information doesn't necessarily address is how the technology could have monumental significance for the whole field of agriculture. What if we could grow *all* our food under cover? That sounds preposterous because we are accustomed to thinking of farms as rather large acreages. But especially with climate change in the picture now, the idea is not so crazy. To put an area of, say, two hundred acres under plastic would indeed be prohibitive for one farmer. But on the typical garden farm of the future, a continuing trend toward more enclosure agriculture looks plausible. Think of the fact that the United States has more acreage in lawn than in cultivated crops. Let us say that fifty million homes (the United States has a population of around 335 million right now) would each put an eighth of an acre under cover. That would add up to some 6,250,000 acres. With the higher yields possible in enclosed farming, each of those acres might produce three times the yield of an open-air acre, or the equivalent of 18,750,000 acres, right? (If that sounds dubious to you, read Eliot Coleman's books, where he describes how to get five crops per year under cover in Maine.) Total cropland in the United States is right around 442 million acres (and, surprisingly, the number is falling). If each of those enclosed one-eighth acre "farms" were housing six egg-laying and meat-producing hens and a pig or two to eat the plant parts from the greenhouses too overripe for humans, we could be looking at a significant amount of food production that would not depend on the gambling whimsy of the Chicago Board of Trade or the weather.

Enclosed farming makes saving rainwater for irrigation purposes more practical. All that water running off the high tunnel and low tunnel surfaces (as well as off the roofs of urban houses and buildings) could be gathered in cisterns and tanks and doled out to the enclosed crops rather than causing flash flooding in the streets. Rain barrels are becoming sort of the new symbol of green energy for this reason (and an excellent example of natural decentralization). It does not take a genius to conclude that this means urban areas are as practical for intensive agriculture as rural areas.

Another advantage of enclosed farming is protection against wild animals, birds, and predatory insects. The United States is becoming overrun or repopulated, depending on your point of view, by wildlife. So far gardeners and farmers seem to be the only ones who realize how critical this problem is because society at large has been brought up thinking that wildlife is endangered. A few species are, of course, but many others are on a rampage. Power outages caused by squirrels chewing electric lines and shorting out transformers, for example, are reaching epidemic numbers. The only reason this problem has not become monumental is that the squirrel gets incinerated in the process, never to chew again. If I were to start incinerating squirrels with a flamethrower, think of the hue and cry that would be raised. But it's okay if our electrical grid dispatches the cute little bushy-tails by the thousands. Just look the other way. Even PETA members need electricity more than they need squirrels.

All sorts of ingenious examples of enclosure agriculture that keep animals out of the garden keep coming on the market. Plastic screening and protective row covers are available now that are not prohibitive in cost. They not only protect against some insects and birds but can shade out weeds and preserve moisture. In our gardens, Carol and I have learned that it is relatively inexpensive and quick to unroll light plastic netting over strawberry beds to keep deer from grazing them. To keep deer out of other crops, we have perfected our low-cost "tightwad" fence. It consists of setting posts in the ground about ten feet apart and about six feet high. Fencing should really be eight feet high to keep the deer from jumping over, but we have found that we can slide sections of old livestock fencing of five feet in height over the posts, leaving the fence a foot above the ground at the bottom and sticking up six feet or so above the ground. The deer could jump over, I suppose, but so far they haven't. At the bottom, we install chicken wire under the livestock fence to keep out the rabbits.

Jan Dawson and Andy Reinhart of Jandy's market garden, whom I turn to often as good examples of what is happening, installed their first high-tunnel hoophouses about six years ago and say now that they could hardly stay in business without them. They also caution me that hoophouse farming is not foolproof. "When you enclose nature by creating an artificial environment, you open your operation to about as many problems as you avoid," Jan points out. "The trick is to be able to open up the greenhouse easily when conditions call for it. There are now high tunnels with covers that are fairly easy to roll up and roll back down as necessary." Enclosure farmers have also improvised ways to move their hoophouses on wheels and similar wheels are now available from garden supply businesses. That way enclosure farming can be rotated from one plot to another.

Improvements in the tunnel and row cover systems are being made constantly, as you would expect: handier ways to utilize the water off the roofs for irrigation; better ways to catch and store sunlight for additional heat on colder nights; quicker ways to handle plastic mulches, including machines that can "eat up" old mulch for easier disposal.

Enclosure farming allows the farmer to experiment with food plants not acclimated to the area's weather. Tea is being grown in hoophouses even in more northerly areas. Jandy's has been experimenting with ginger roots to sell at market along with their more common vegetables.

The hoophouse technology is being used as a cheaper way to raise relatively small numbers of farm animals, too, especially in conjunction with deep bedding, as I alluded to earlier. The manure, instead of being a liability as in most factory production, is absorbed into cornstalk waste used for bedding as much as four-feet deep. Hogs root through the bedding, eating some of it, and stir the rest into excellent compost, while the heat coming through the roof along with the heat of the compost keeps young animals warm. As the prices of chemical fertilizers rise, the manure and bedding become almost as valuable as the pork. The deep bedding soaks up the liquid manure so that expensive storage tanks or lagoons aren't necessary.

One of the most successful practitioners of enclosed farming is Eliot Coleman, cited earlier. He grows vegetables all year, winter and summer both, in Maine. He manages this by really radical enclosure farming using mainly movable greenhouses and the humble, old cold frame. In the November, 2014, issue of *Acres U.S.A.*, in an exceedingly interesting interview ("Extending the Growing Season—Organic Farmer, Author Eliot Coleman Shares Strategies for Successful Year-Round Growing" by Chris Walters), he says:

As we continued experimenting with that in the '80s, we put cold-frames inside a homemade greenhouse, and all of a sudden that was a great leap forward.

It turned out that each layer of covering moved the covered area climatically about 500 miles to the south. So outside I'm in Maine. I walk into the greenhouse, the first layer of covering, and I'm in a climate like New Jersey. If we have a cold frame in there, I reach my hand into the second layer of covering and my hand is in the Georgia winter climate. Obviously there are no tomatoes or peppers or eggplants in there, but there's plenty of spinach, carrots, scallions, Swiss chard, kale and Asian greens that don't mind freezing at night as long as they're safe from the dry, desiccating cold winds.

There is another way to look at this whole topic of decentralized food production. How about some real heresy? Maybe farming isn't an economic process that is supposed to make money. What if food production should be an activity that nearly everyone has to be involved in, like cooking or taking a bath?

The first commandment of agriculture states that you must put back into the soil the fertility you take out of it. If that admonition is truly followed, where is the profit, other than the joy of eating and thereby sustaining health and well-being? Any actual money profit beyond that might simply be a sign that the farming is flawed. Failed civilization on top of failed civilization suggests that this might be the case.

Turning agriculture into a business that sells industrialized commodities doesn't work very well anyway, which is why farming has so often been subsidized. A commodity that fits into industrialized manufacturing is something that people make in a shop or factory, where nature can be sort of controlled. Crops are not really commodities. They grow in the natural world and grow at their own sweet, natural pace, one that is not easily adjusted to supply and demand, much less artificial money interest. GMO scientists crow about their new seeds, which they assume will lead to a new era of profitability for agriculture. But in fact, there have been documented decreases in yields. Most above-average increases come from good weather and good soil practices as old as Adam. Monsanto, Syngenta, Bayer, and others try to take credit for big corn crops when their very same seeds that produce a good crop on one farm result in only half a crop two miles down the road where timely rains did not fall. And even when the quantity of yields increases, quality sometimes decreases. Moreover, the "good" management practices of the high-profit farmers are often bad practices in terms of extracting wealth from the soil (in the form of mineral nutrients and organic matter) and not returning enough of it to balance the equation. Some poor to average farmers in terms of yields could be rated as the best farmers, environmentally speaking, because they are mining the soil the least.

The very naturalness of food gives small-scale production another advantage. Critics often tell me that my idea of feeding the world with small-scale garden farms is naïve, like saying we could supply everyone a homemade car. But a car requires all sorts of overhead money and artificial outside inputs, while food grows mostly on the ubiquitous gifts of sunlight, soil fertility, and rain.

And while I'm talking crazy, what if there were a hundred million people in the United States raising a half-acre of corn as a hobby? That would add up to enough corn for all our food needs and no one would make monetary profit doing it. That would really be a movement to match the bowel kind, enough to scare the shit out of the moneychangers. And wouldn't it make at least as much economic sense as maybe a couple hundred thousand farmers spending billions to grow corn and still having to be kept afloat by government subsidies and borrowed money?

I try, in my mind, to find a way to describe without sounding naïve, in a world surfeited with money, a kind of garden farming that is more rewarding than "for-profit" farming and that still contributes to a healthful food supply. All of us, except the richest, must get some money from our work. How much is "some"? The closest I have come to describing garden farming the way I live it goes like this. We keep a few chickens because we like our own fresh eggs, our own fried chicken, and we can always find uses for the manure and bedding (could also use the feathers to stuff pillows, come to think of it). Also, I enjoy the company of my almost pet hens. We have surplus eggs to sell, and I appreciate getting paid for them. But I would give the eggs away and do not do so only because the people I want to give them to absolutely insist on paying for them. We charge less than market price so they save a bit anyway. We both benefit because health, happiness, and friendship are the goals. Is this capitalism, socialism, distributism, or something that has not yet been named? Both our customers and ourselves are "profiting" from the eggs, but it does not show up as profit or loss on anybody's accounting ledger.

The business response to my argument would be that I can "afford" to give away eggs or sell them cheap because I make my money not from hens but from hen-scratching on computer screens. (Scratch, hen, and live.) My rebuttal is that I can only "afford" to keep on writing because of my "non-profit" eggs, fried chicken, and other garden farm food, fuel wood, and recreation. Some writers make a lot of money, just as some farmers do, and that's okay. But is it not also okay for many of us to do what we like to do so much that we are willing to be content with a lower-middle-class income? What would I do with real money profit? Buy a bigger or newer car I don't want or need? Build a bigger house I don't want or need? Buy more land? I probably would have done that in my younger years, but fortunately I was saved because I did not have the money. If I had bought more land, using borrowed money, it would have distracted me from my writing. And, knowing me, I might probably have lost it all by now.

Another example: For years some of us have written against concentration of huge numbers of chickens or livestock into giant factories. Our main reason was that if disease struck the industry or instability wrecked the society, it would be far better to have many small farms than a few giant ones from which to draw food. We lost that debate because, we were told, the only way to produce eggs profitably was in huge numbers. Look now what is happening. The bird flu epidemic forced factory farms to kill their chickens and turkeys. Millions of them. Some of the growers faced bankruptcy. The price of eggs doubled. Were it not for thousands of backyarders keeping a few hens, the disaster would have been even worse. Wouldn't it be better if at least half the egg production could come from backyarders?

Surely it is possible to live life free from the shackles of paper money growth. At least partially free. I see the dream expressed in one new garden farming book after another, like *The Nourishing Homestead* by Ben Hewitt with Penny Hewitt, and *The Lean Farm* by Ben Hartman (both published by Chelsea Green in 2015). I see it proclaimed in countless blogs and in newer farm magazines. The language of artisanal farming is the same language as that of all the other arts. It is nice to make a lot of money from a book, or a painting, or a song, but that is not why real artists do it. If they can just make a modest living from their work, they are satisfied. If they can't, they still keep on working on their art and do something else for necessary income. New garden farmers are the same. They talk about soil enrichment just like artists talk about paint. It is not about money but about beauty, inspiration, taste, and timelessness.

Lasting forever is the goal of all art, and the very soul of good farming. Money can add a little sizzle to it, but it's more a distraction than a goal. The real reward is being able to look out over our gardens and farms and know that we are adding real value and beauty to the world, or at least doing no harm. All is right in our world. Except maybe that bull thistle I see sticking above the pasture grass out there. Where's my hoe?

CHAPTER 4

The Ripening of a "Rurban" Culture

I n the last hundred years, the "flight" of people from farms to the city has gotten much attention, but starting about 1960, there was also a "flight" of urbanites into new suburban developments and what is sometimes called exurban communities beyond the suburbs. In our very rural county, for instance, a significant number of farm children who moved to town to get a job have been buying acreages off the old home farms and establishing their new homes there. Much of the landscape of the United States is now at the point where a traveler has a hard time discerning where a town ends and farms begin. Geographers and sociologists call it a "rurban" society.

When I was a child on the farm, I took some pleasure in repeating a remark that I now realize was quite prejudicial. An old farmer in our neighborhood, rankled because he did not like people from town building houses on farmland, would growl on all occasions: "Them that works in town should live in town and them that works in the country should live in the country." My siblings and I would mimic him, incorrect grammar and all, and giggle. We knew what he said was more than a little over the top, but it reinforced our anger at being made fun of in school because we were "country hicks" with imagined traces of manure on our shoes. So we called them "city slickers" and pointed out that as soon as they got enough money, they wanted to move out into the country. Yes, old neighbor: "Them that works in town should live in town." Of course we adroitly avoided facing up to the fact that many of "them that works in the country" went to town to live it up whenever they could find an excuse to do so. Truth is that rural and urban cultures have been merging since the end of the hunting and gathering age, but we like to pretend differently when it suits our purposes.

That kind of cultural merging is about to become complete, I think. The economists have predicted the demise of small-scale farming for at least sixty years because, they insist, it just can't compete in the real money world (as if the money world were real). But many contrary farmers of the garden variety belong to the Not Necessarily So Society, and like the bumblebee that isn't supposed to be able to fly, they just buzz happily along from one clover blossom to another, managing to avoid contact with OBD (Over-Borrowing Disease), which is the reason so many of their brethren had to sell out and move into town. There is an increasing number of garden farmers now who know how to make at least a modest living from just a few acres. And there are two-hundred-acre farmers who are farming quite profitably with horses, which must give the proponents of OBD the fits.

The most intriguing fact about this new garden farming society is that it is happening more in and around cities than out where the tall corn grows and the big tractors rumble. This surprises some farm observers. What do those city slickers know about farming that we real farmers don't know better? But this primacy of the city in agricultural development has always been the case. Odd as it seems, agricultural innovation almost always begins in town. My favorite mind-stretching book, *The Economy of Cities* by Jane Jacobs, documents the historical evidence in favor of this conclusion so well that it is almost impossible to refute her, though before I read her book, I tried. I was still under the influence of "them that works in town should live in town, etc." I didn't like the idea of urban culture dictating farming methods, but it was all too painfully the truth. "New kinds of farming come out of cities," Jacobs writes. She gives numerous examples. "The growing of hybrid corn . . . was not developed on corn farms by farmers but by scientists in plant laboratories, promoted and publicized by plant scientists and editors of agricultural papers, and they had a hard time persuading farmers to try the unprepossessing-looking hybrid seeds." In another instance she points out that when the wheat farmers of New York realized they could no longer compete with Great Plains wheat growers [or thought they couldn't] and switched to fruit farming, the change came primarily from "the proprietors of a nursery that first supplied the city people with fruit trees, grape vines, and berry bushes and then showed farmers of the Genesee Valley . . . that orchards and vineyards were economical alternatives [to wheat]." Likewise, "the fruit and vegetable industries of California did not 'evolve' from that state's older wheat fields and animal pastures. Rather the new California farming was organized in San Francisco for supplying fruits to preserving plants and later vegetables to vegetable canneries."

The fact that cities inspire agricultural development proves to be true as far back as we can go in historical and pre-historical records. Where established farming activity flourished, it was always in conjunction with established communities of people. Hunters and gatherers more or less alone out in the boondocks lived on wild food. After they congregated into cities, not enough wild plants and animals were easily available, so gardening and husbandry began on adjacent land. All of the farmers of that era were garden farmers. Alfalfa was a medicinal garden plant long before it became a hay crop. Cornfields were as much a part of prehistoric cities in America as burial and effigy mounds. Even among historical Native Americans, cornfields were associated with villages and towns.

In the earliest times, communities would plant a plot to grain and, when the soil lost its fertility, would switch to another plot.

Then they noticed that plants grew better around their midden pits of food wastes and bodily excrement. Fertilizer! South American prehistoric cities went on to develop what we now call "biochar"—definitely a fertilizer that required concentrations of human population to become practical. Today, fertilizer discoveries keep coming out of cities. For instance, refining phosphorus fertilizer from treated sewage sludge looks promising.

The modern organic farm movement began in urban culture, most particularly in Emmaus, Pennsylvania, where J. I. Rodale, fresh out of New York City, bought a farm. When he started his magazine, *Organic Gardening,* he sent out fourteen thousand copies free to farmers hoping to stir up interest. His son, Bob, once told me, with that characteristic whisper of a smile on his face, that his father got back only twelve subscriptions. Commercial farmers resisted organic methods. When I moved back from the city to the farm country of my boyhood to write for organic-minded publications, some farmers in the neighborhood wondered, only half-jokingly, if I had sold out to the "pinko-commies."

A great example of how city and farm are parts of the whole occurred in and around cities in the era when horses were the main form of transportation. Dire predictions were appearing in newspapers in London and New York about how the streets in fifty more years would be six feet deep in horse manure. The piston engine removed that fear, but before that happened a marvelous urban agriculture blossomed, ringing cities with food gardens that used substantial amounts of horse manure as fertilizer and for heating the hotbeds, which produced out-of-season fruits and vegetables in protected culture. Today while exhaust from piston engines threatens to bury our cities in a six-foot layer of carbon monoxide, urban garden farming is on the increase again, using composted leaves and other organic wastes for growing fruits and vegetables in gardens and hotbeds and on rooftops and using all those roofs to gather irrigation water. As the environmental problems increase on the West Coast, where so many of our fruits and vegetables originate, especially the out-of-season supply, this kind of urban farming could become a life-saver in the future and prevent or offset who knows how much piston engine pollution.

The notion that cities and agriculture proceeded hand in hand is perfectly logical if you think about it a little. Agriculture blossomed where people congregated together because of the ancient wisdom of "many hands make light the task" and "two heads are better than one." Humans are social animals, much more disposed to having other people around to enjoy (or fight) and for mutual aid. Historically, in many parts of the world, the more traditional way of life for farmers was (and is) to live in town and go out daily to their farms round about, returning in the evening to enjoy each other's company and trade experiences. That's how new ideas arose on how to improve crops and invent tools to make the cultivating and harvesting easier. The blacksmiths who first made those tools were almost always "village" smithies, not farm smithies, and their shops became factories, and the factories became John Deere and International Harvester. As it has worked out, "them that work in town" often live in the country and "them that work in the country" often live in town, and thus it shall always be.

Jane Jacobs gives an appropriate example of how easy it is to believe that settled farming came before cities. Electric generating plants are often built out in the countryside. If future archeologists found ruins of one of them out in the country, the old way of thinking (them that works in town, and so on) would deduce that farmers had originally developed electricity and then, when a surplus developed, cities became possible.

The new garden farming movement is being driven by upscale restaurant chefs, by the flocks of people who frequent farmers' markets looking for fresher food, by a growing number of people who are suspicious of regular commercial food, and most of all by a millennial generation that seeks a different philosophy to live by than what industrialism offers them. Behind it all, or because of it all, a demographic change that has been going on for a long time is reaching maturity. The whole notion of city "limits" is archaic. Rural culture and urban culture are fusing, raising havoc with zoning ordinances and neighborhood relationships. People are screaming at each other over so-called ill-kept lawns, backyard chickens, loose-running dogs, horses pooping on roads, speeding cars, tractor noise at night, and a thousand other little grievances that are the growing pains of acculturation. The problems will be mostly resolved once everyone understands how cities and farms are parts of a whole, not divisible one from another. When we all realize that, as we munch our good, fresh food, it will not only mean a better environment for all, but the end to this silly political anger that colors everything blue or red instead of a lovely productive green.

Another important facilitator to the obliteration of boundaries between rural and urban life is electronic communication. The computer is making us all more or less one culture, even while it does its best to twitter away at keeping us divided. One very small example: I read the Sunday *New York Times* and the Cleveland *Plain Dealer*, which (again proving the point) sometimes carry more insightful articles about innovative farming than some farm magazines do. But it is becoming more and more difficult to get them delivered to my doorstep out here where the tall corn grows. Doesn't matter. As I wander across my pasture, I can read them on a smart phone (or could if only I knew how to do the proper button-clicking).

But the whole new agriculture idea would stumble and falter were it not for something else in the wind, and in the soil. The new garden farmers do not go into this business with the idea of making a lot of money. They don't want to go broke either, of course, but they are drawn to this kind of artisanal food production for reasons other than how much money they can make at it. Older, traditional farm magazines, like the one I worked for years ago, liked to talk about farming as a way of life, but they talked out of both sides of their pages. Almost all the articles held up the allure of highest possible profits as the goal. Their names gave them away: *Successful Farming, Top Farmer, Top Operator* (the one I wrote for), *Executive Farmer, Profit Farming, Progressive Farmer*. Never was there a *Slowpoke Farmer* or a *Laid-Back Farming*. The ideal was for every farmer to succeed monetarily, and however one wants to debate the worthiness of that goal, the fact is that it rings hollow. In my lifetime, the number of commercial farmers has decreased from some nine million to roughly five-hundred thousand of which less than three-hundred thousand do most of the heavy lifting today. The notion that getting ahead in farming means making a lot of money is historically a pitiless lie for the vast majority of people.

We are past that, I hope. I don't have statistics to prove it, but I'm fairly sure that the most popular and successful "farm" magazines today are the *Edible* publications mentioned previously. There are eighty-three of them so far, serving various parts of the United States. The articles feature farmers, gardeners, restaurant chefs, food wholesalers and retailers, and the hundreds of thousands of people interested in good, wholesome, high-quality food. I combed through a whole issue of *Edible Ohio Valley* and could not find in any of the articles about farming and small food businesses the words "higher profits." Instead of profitability, the key words throughout were quality and sustainability. In one article, on vineyards and grape growing, the viticulturist is quoted as saying flat out that, in her business, "you're not going to be rolling in money."

CHAPTER 5

~

The Barns at the Center of the Garden Farm Universe

The traditional barn, or something like it, is returning to the scene as more people get interested in garden farm husbandry and aim for quality, not quantity, of product. They are looking at how these barns of yesteryear were designed for comfort and for low energy costs. Some are remodeling the old ones instead of tearing them down or burning them, as has been the habit. Others are building new ones modeled on the old. Their beauty alone is worth the extra cost.

During lambing time last spring, my sister, Berny, and her husband, Brad, invited me to spend some time with them and their sheep in the barn on the home place. I don't know if I've ever experienced a more peaceful and magical afternoon. Those calm ewes, some munching hay, some sleeping, and their lambs bouncing around or nursing, with clean, dry bedding underfoot, exuded good husbandry. At one end of the big open area, Brad had partitioned off a stable for their draft horse, who has been known to round the sheep up in the pasture and drive them to the barn as proficiently as any border collie. As I sat there remembering the past, Brad and Berny stayed busy moving expectant mothers into pens or playing midwife to birthing ewes, all quieter than a hospital, casting a spell of tranquility over the barn that was palpable and soothing. This barn used to be our hog house when I was a child. Brad built hay mangers especially designed so the sheep can't waste hay by pulling it out and dropping it on the floor. He installed a loft overhead to make it easy to drop hay into the mangers. The quiet among the ewes and lambs and horse seemed unearthly, but actually was very earthy. It's hard to put the calm feeling into words. It was sort of like being in a church in the middle of the day when no one else is there, a certain muffled quietude that is calmly unworldly. In fact for some farmers their barn is their church.

Even when the livestock are all out on pasture on a summer day, there is a dark, cool, silence in the barn enhanced perhaps by pigeons cooing up near the roof and the hens singing in the barn-yard or clucking as they scratch in the cow and horse manure for fly larvae inside the barn. It is almost pleasant to be forking manure into a spreader at this time. The secret is never to be in a hurry, to let nature's pace rule. Fork a little. Take a breath. Discuss the news with the chickens. At certain times of my life, I milked cows in the evening after supper, the customary time on the traditional farm. It was even more magical then, milking by moonlight (see N. C. Wyeth's painting *The Springhouse*). The cows were calmer then as twilight turned to darkness and the moon peeked through the barn door. So were the farmers, and they did not usually have evening appointments elsewhere to hurry them along. Their life at that time of day was right there in the barn, full of contented accomplishment.

Sitting in a quiet barn, surrounded by your animals, you feel you are in a citadel of security, safe from the onslaughts of money markets, weather, and human madness. There is an independence here and a pride that gives a person a kind of satisfaction not common anymore. If the electricity goes off, you might not even know until it comes back on again.

The horse lovers' barns, often more extravagant than need be, especially provide evidence that it is okay to farm more for comfort than for "efficiency," even if it does not make any real profit or makes just enough to allow one to continue his or her chosen vocation. Let's face it. If you keep riding horses or farm with horses, you are not aiming for a couple thousand acres and a bank directorship but are happy with a couple hundred acres or less because, among other things, your horses return the love you show them, are generally cheaper than tractors, and they always start in the morning. Moreover, one never knows when something you decide to raise that seems impractical and unprofitable suddenly shows commercial possibilities. Today, an Amish farmer sometimes can sell a workhorse to wealthy breeders for more money than he made farming all of last year. Whoever thought that raw milk would find a viable new market? Or that eggs would again become the darlings of the food faddists, much less butter and cream? Who would have believed that a consortium of very small beef and pork producers, along with the small farms that supply them with young livestock and the local small slaughterhouses and butcher shops that process the meat, would generate a considerable amount of money, all totally outside the pale of industrial factory farming? Just down the road from us, a roadside marketer and his wife rebuilt the old barn on their property, not only to use as part of their store but to house their workhorses. The barn and the horses both attract customers. How do you figure the profit from that?

While the way you arrange your pens and aisles and lofts will be based on your specific needs and desires, there is a wealth of traditional barn-architecture lore that can save you time and money. When I built our barn to house a few sheep, a horse, and two cows, I studied the angle of roof to wall in traditional barns, wondering if maybe there was a standard commonly followed. There was not, as far as I could find, but by using a traditional design where the roof in two sections rises steeply from the first floor to the roof peak, I could get almost as much space on the second floor as if I had a third floor. (Not thinking so few livestock merited buying a baler for making hay, I planned to store hay loose rather than baled and so needed more loft space.) I also learned that an ear corn crib's narrow, four-foot spacing between walls and the slant of the walls outward from bottom to top were design details steeped in traditional experience. The spacing between walls should not be more than four feet to allow for effective natural air penetration to dry ear corn. The outwardly slanted walls allow moisture to drip down outside the crib not into the corn.

Another good way to learn the sophisticated and labor-saving techniques of traditional barns is to study old books on barns like those by Eric Sloane. Better yet, visit museums dedicated to agricultural history. One of my favorites is in Heritage Park on the campus of Otterbein University in the Columbus, Ohio, area. The barn on display has a three-story wooden tower at one end, with a windmill sticking out the top. In earlier days, the windmill pumped water into a large storage tank on the top floor. The water then fed by gravity down to the first floor for the livestock to drink or for whatever other uses the farmer might have for it. Really cool.

The barn magic that is following the new garden farming extends to possibilities much greater than it at first appears. For example, a reliable source tells me about some Amish boys who found a new way to make a little money from their old barn. It was furnished with dovecotes, as many traditional barns are, to attract barn pigeons. Pigeons, or rock doves, have from time immemorial been part of the traditional farm, providing delicious pigeon pot pie without much cost while taking advantage of the fly control which pigeons can contribute around the barn. And if you check Google, there are all kinds of sites selling squab. There is also the possibility of raising and selling homing pigeons. The Amish boys learned that a nearby game farm was interested in buying pigeons for great white weekend hunters to shoot at. Great white weekend hunters usually miss, and the pigeons, endowed with a bit of homing instinct, fly back to their native barn, where the boys capture and sell them again. How's that for "per unit efficiency"?

Part of the wisdom built into traditional barns is their use of natural efficiency, some of it only now coming to light. Considerable research in medical science is suggesting that children who spend a lot of time in barns with farm animals have a significantly lower chance of developing asthma. What's that worth? As for farm work itself, you are hardly ever doing just one thing when you are working in the barn. When you are feeding and bedding down the animals, you are making the fertilizer for next year's crops. And oh, how the cows love to romp in the fresh straw you put down for them to sleep on. Happy cows give more milk. The clover hay you feed has already partly paid for itself by supplying free nitrogen to your soil. The hay also often has enough protein in it so you don't have to buy protein supplements. The hay piled high in the mow is a monument to its role in controlling erosion when it was growing in the field. Nothing is wasted. Half-digested corn in the cow manure becomes food for the chickens as they scratch in the bedding for fly larvae. The cows keep the barn warm enough so the old-time water tank overhead, kept full by water from the roof, doesn't freeze. Where appropriate, windmills can be positioned to pump water to cisterns on a higher elevation than the barn, and the water flows by gravity to the watering troughs and tanks inside the barn. In a bank barn, where the structure is built into a hillside or embankment and only one side is exposed to the weather on the bottom floor, I used to milk coatless in 0° Minnesota weather. And in summer the half-underground space stayed cooler. Later, milking in a more modern barn, we had to install a stove in winter and fans in summer to keep us comfortable.

Barns promote wildlife. When I built mine, I unintentionally provided housing for bats between the plywood plates that hold the rafters together at the peak of the roof. Seldom is my barnyard in the woods clouded with mosquitoes. New research, recently reported in *Acres U.S.A.* magazine, suggests that bats perform significant control of corn earworm by preying on the earworm moths. What's that worth? Many new farmers also find their barns attract barn swallows and barn owls. Mine attracts robins and phoebes. It also attracts pesky raccoons, but some years their pelts can mean extra cash, too.

The most peaceful, pleasant time in the barn is in winter. You might have to wade through snow or bow to biting winds to get there, but once in the barn, how quiet and warmish it is. The atmosphere is especially comforting when you need to stay in the barn at night watching over birthing animals. Animal heat and the insulation of the haymows keep it cozy. The composting bedding and manure keeps the floor warmish too so the animals can sleep there comfortably. I think it reduces mastitis in cows. There are no flies in the winter barn. The sheep make little gurgling noises of contentment as they nose into their evening meal. Cows and horses crunch away, making that hay sound almost good enough to eat myself.

The work is challenging but pleasantly so. Numbskulls do not last long in the barn. You know most or all of your animals by name. Curlyhead needs to be penned tonight because she is about to have lambs. You learn how to play midwife if necessary. Make sure the rooster, Trump, is in the coop before you close up, or he will be crowing in your bedroom window at daybreak. Give Shorthorn, the steer, a little more grain tonight, and don't let Whiteface push him away from it. While the calf suckles on one side of the udder, you can milk your share from the teats on the other side. Make sure that little pigs and lambs are all back with their mothers before you go to the house. Be sure to turn off the hose filling the watering trough if you don't have automatic waterers. The traditional barn I grew up in, and which was common in our area, had a marvelous "automatic" watering system. Next to the windmill by the house was a big wooden tank made of cypress wood that, when moisture swelled the boards, did not leak. It was open at the top but did not freeze over more than eight inches and usually less than that. The windmill kept it sufficiently full of water. Pipes ran out of the bottom to the house and barn buildings to supply water, again by gravity, to small cypress tanks in the barn lots and to the cistern under the house. Floats closed the pipes when the waterers were full.

Traditional barns use gravity for power whenever possible to avoid a lot of heavy lifting. That's another way bank barns bring comfort to the farmer. Animals enter on the bottom floor open to the outside. Entrance to the upper floor is on the other high side of the bank. I know one farm in Pennsylvania built into a steep hill where there are ground-floor entrances to all three floors. This makes filling haylofts and granaries as well as dropping feed down chutes and hay mow openings comparatively easy.

Farm barns are menageries where the animals become pets of their human caretakers and especially of the children. Also, for children, barns are big playhouses, a place for games of hide and seek, of swinging on hay ropes, of sliding down mounds of hay. The barn dance is still part of our cultural heritage. Basketball historically seems to have started in barns. As the mows empty and warmer weather arrives, the floor space opens up. Some say that's why basketball's main season is spring. And many barns are still used that way—ask our grandsons.

Nor need barns be limited to traditional design to achieve some comfort. There are new structural materials that fit the small-scale, artisanal food economy very well. As I pointed out earlier, hoophouses, covered with fabrics of various kinds, make low-cost greenhouses or livestock sheds that are essential to new market gardening. Such structures, along with deep-bedding techniques that generate heat, are being used for sows and pigs, too. These buildings essentially are all roof, and the sun coming through helps provide warmth during the daytime. The husband-man or market gardener can lay out the floor to suit his needs because there are no internal structural supports encumbering the space. It is interesting to read the first line of advertising about these hoophouses: ". . . provides a stress-free, healthy environment for both livestock and workers."

You will invariably start talking to your livestock while you work with them. And I don't mean just directional commands. I find myself discussing religion and politics with my sheep after hearing the latest absurdities on the radio. Sometimes the sheep talk back. "Did you hear *that*," I exclaim to Curlyhead. "That guy just said that livestock are one of the biggest threats to our environment!" She just keeps on placidly chewing her cud and replies, "Baaaah." Sometimes neighbors or friends come into the barn unannounced as I carry on discussions with the animals. If they are farmers, they laugh understandingly. If not, they wonder exceedingly about my sanity.

As you loiter after chores, you think about how you and your barn sanctuary form a sort of halfway house between man and nature. You are filled with great satisfaction and a feeling of independence. The world beyond might be foundering in chaos, but right here in your quiet barn, peace and sanity prevail.

CHAPTER 6

Backyard Sheep

"Stop mowing and start growing" is the motto and battle cry of a fairly new (2011) organization called Urban Shepherds. Its purpose is to encourage grazing sheep on urban and suburban vacant lots, larger lawns, and other grassy areas like school campuses and the acreages surrounding historic sites and factories. This particular effort originated at Spicy Lamb Farm near Peninsula, Ohio. The farm schedules sessions regularly to inform people how sheep might just follow chickens as a practical back-yard addition, especially on country estates and larger open spaces kept in grass. Spicy Lamb Farm already uses a power line right-of-way close by to graze its sheep, benefitting the power company and itself, controlling weeds without mowing and without using nearly as much herbicidal spraying that would otherwise be necessary. Cleveland, Detroit, and Akron all have Urban Shepherds projects under way, and other parts of the country are catching on to this idea. One of the selling points is that if you have a business open to the public, grazing sheep have proven to be a big attraction.

All well and good. Why should chickens have all the rights to the backyard barnyards of America? I like to think I had something to do with this laudable project, at least indirectly. When our daughter and her family moved to the Cleveland area years ago, I had a chance to observe suburban lawns closely. I was amazed how they stayed green most of the winter. Many times, trying to be funny, I wrote that suburbanites were the best pasture farmers in America and didn't know it.

I doubt that many lawns will actually turn into sheep pastures in the near future. There are too many people who will have conniptions at the idea of sheep next door fertilizing the grass with that awful, vile stuff called manure. As if the deer and rabbits traipsing over the grass didn't defecate. Most suburbs have regulations against fences, too, so that would be another hurdle that would have to be overcome. But larger estates and all kinds of grassed areas around schools, historic sites, factories, parks, and public utility rights-of-way could work. I know a country cemetery (near Hepburn, Ohio) that has used sheep to keep down the grass. The grazing animals are particularly advantageous in this situation because they can bite the grass off right next to the tombstones. Historically, golf courses used sheep to keep down the grass, and there's talk of doing that again.

I suppose someone will soon market a diaper for sheep. Diapers for chickens are already a fact. We used regular baby diapers on bottle lambs we kept in the house when they were first born. But outdoors, sheep manure—mostly little pebbly turds just like what the deer and rabbits excrete—is really not offensive or an odor problem if you have only a couple animals. The manure just disappears down into the grass to become fertilizer, saving the homeowner that expense.

I salute the Urban Shepherds and wish them luck. Their efforts will at least get more new farmers interested in raising sheep on pastures and selling "free-range" lamb, which is more practical than free-range beef because sheep are easier to manage than cows. It could also help another advancement in civilization. There are fathers and mothers now who adhere to what they call "free-range parenting." They are rebelling against helicopter parenting, I guess. They think their children need more freedom during playtime, instead of being constantly corralled with too much oversight and regulation.

The garden farm is the perfect solution to this problem. It is time for free-range grazing to join hands with free-range parenting. Turn all that wasted suburban yardage into home-on-the-range and teach children what the real world is like by exposing them to shepherding. They can spend delightful hours, while fingering their smart phones, making sure the sheep have water, guarding them against dogs and coyotes, and spotting breaks in the fence needing repair. Then the ancient nursery rhyme would come full circle. "Where's the little boy who looks after the sheep? / He's out in the haystack, fast asleep."

Pay attention now because, if you want to raise sheep to make some money, or merely to keep the lawn mowed and get a few racks of lamb in payment, I am about to save your life.

We generally refer to male sheep as *bucks* in our neck of the woods, but *ram* is probably a better term since everyone here in Ohio thinks bucks are football players at Ohio State. In any case, the ram or buck of the woolly kingdom has taught shepherd and shepherdess alike the perfect literal and metaphysical meaning of contrariness. He may be standing there in the shade of his shed, seemingly at total peace with the world, barely even deigning to look at you as you pass by. But turn your back and he will plant his head into your butt and send you to the nearest chiropractor for the rest of your life. And don't think you can teach him a lesson by breaking a cane over his skull. Rams love getting hit, especially in the head. I think it gives them orgasms. The only way you can get any respect is to rap them sharply on the nose with a short, stout stick that you should carry in your pocket whenever you are within a half-mile of them.

When I hear an animal lover who has never had to take daily care of animals criticize the way we husbandmen and husband-women treat our livestock, I wish that they had to learn reality the ram way. I look with considerable reservations at all those sweet biblical pictures of "good shepherds" who leave the ninety-nine behind to go search for the one that is lost. Why are there no pictures of good shepherds getting nailed in the butt by a ram—a scene a whole lot more common? Besides, sheep are never lost. Shepherds just can't always find them.

For some reason, in agrarian cultures, nothing is as funny as seeing a ram send a farm boy flying into a pile of manure. It has happened to all of us who raise sheep, so maybe it is just a matter of misery loving company. I don't care how carefully you keep an eye out—the moment you forget and turn your back, BAM. Most of the time no harm is done, which I suppose is why it seems so comical (especially if it happens to Dad after he has scolded you for something your sister did). But ram attacks are not funny. Rams can kill.

Do not try to run away from an attacking ram. That is suicide. No matter how bedraggled or decrepit he looks, he can outrun you. To survive you have to study contrariness in the flesh. If you watch two rams about to deliver orgasms to each other, they will face off and take a few steps backward. Then they charge, colliding head on with enough collective force to make an anvil bleed. Then they quiver with pleasure and do it again.

So when you see your ram start to back away from you, that's the dead giveaway (*dead* being the appropriate word here) that he is about to kill you. Walk towards him. No matter how suicidal that may seem, walk towards him. I mean go right at him. Almost always this is confusing to a ram and he will keep backing away for awhile and might lose interest in killing you. This can give you time to get closer to a fence or a tractor. If you can get to an immovable object like a tree, all you have to do is keep it between you and the ram. Then he can't do his classic charge and soon tires of the game.

Otherwise, like out in the middle of a field, he will eventually quit backing up at your advance and attack. Stand your ground. This takes a great deal of nerve the first time. But at the last second before he butts you, he will lift himself onto his hind legs to give his forward motion extra pile-driving force. Up on his hind legs, he can only lunge straight ahead. He can't turn. So when he lunges, all you have to do is step sideways, quickly of course, and his momentum will carry him past you. This maneuver is quite effective, and it is almost comical to see how puzzled the ram will be when all he collides with is thin air. If you are young and strong, this is the moment when you grab him, twist his head around backwards, set him on his ass like you were going to shear him, and pummel the living hell out of him. Pummeling, which I define as slapping from one side of his bullheaded skull to the other with your open hand as hard as you can without breaking your fingers, is a sensation the ram does not enjoy like he does being hammered on his skull. Some shepherds say this kind of treatment will only make him meaner, but in my experience, he will act like a gentleman for about a month. Or will absorb enough fear of the Lord so that when you see him backing up the next time, a warning yell will make him stop short and decide it is more fun to go eat hay.

If you are not young and strong, you should only be out with the flock in the pasture if you are riding a tractor or four-wheeler. I have never tried to challenge a buck with the latter vehicle head on because I'm afraid that it would come off second best. But at least it can go faster than sheep.

One of my brother-in-law Brad's rams, which had also been my ram earlier, absolutely loved to bash his head against anything that moved. When he no longer had a partner ram to amuse him, he challenged Brad's two steers. The otherwise placid bovines took turns bashing him until he finally realized that there was no future in ramming hard-headed bovines three times his size. So (and this is all the evidence you need to prove the insanity of the male hormonal system and to understand why this beast holds such a high place in the contrary farmer hall of fame) Brad's ram went after his draft horse instead. You have to understand that the horse thought he was master of Brad's sheep at that time. So of course ram and draft horse, the icons of contrary farming, were also idiot male rivals. In the beginning, the ram got in maybe two or three good charges before the horse learned to wheel around and blast his attacker into cuckoo land with his hind hooves. I know you will not believe me, but the ram seemed to love getting his head nearly kicked off by flying hooves. He just kept coming back for more. The horse then learned a new strategy. Wheeling all the way around to send the stupid ram head over heels got to be a lot more trouble than it was worth, so then, when the ram charged, the horse elegantly extended one of his front legs and planted his hoof into the hapless ram's lowered head, like a football running back stiff-arming a tackler. That stopped the ram dead in his tracks. Eventually he got tired of tormenting the horse and contented himself with chewing off the end of the horse's tail. You won't find that fact in any book on husbandry.

This all suggests an interesting philosophical question. If I try to cave a ram's head in with a baseball bat, thereby breaking the bat, the well-meaning, civilized human observer would accuse me of cruelty to animals. What if a horse nearly shatters the ram's skull with its hooves? And the ram comes back for more?

Obviously, there are lessons here for anyone wanting to start raising sheep, whether commercially or as lawn mowers. You don't want a ram on your place any more than a rooster if you can avoid it. This also opens up opportunity for commercial shepherds. Renting out rams could become a sideline source of income. So could selling weaned lambs to backyarders who only want a couple. Another possibility is one we occasionally pursued when we had sheep. Orphan, bottle-fed lambs are a pain and not very profitable, but we could almost always find an acquaintance who wanted their children to have the experience of raising a farm animal. Nothing can captivate youngsters like bottle-feeding a lamb. The lamb becomes a loving pet. As I have said, we put a diaper on one and let it have the run of our house. Very entertaining. This can be a way another new shepherd gets started. Or the lamb can become a 4-H fair project. The experience almost always ends in tears when the child must sell the lamb or see it turned into lamb chops, but that also can be a useful lesson in what life is all about.

CHAPTER 7

Hauling Livestock: The Ultimate Test of Your Farming Mettle

In the garden farming age, there are occasions when farm animals have to be moved either to another farm or to market. That's the price we pay for decentralization. Drones might work for parcel post, but hardly for pigs or cows and even then, if it came to that, the animal would still have to be gotten into the drone. Loading farm animals onto trucks is something that requires the patience of Job and often ends up with you, like Job, sitting on a dung heap. Until you can do it successfully and then are crazy enough to keep on doing it, you will hardly make it into the ranks of get-small-and-stay-in husbandry.

I had thought by now that humans new to farming would be smart enough to hire experienced haulers to move their livestock around. But that costs money, of course, and is another business that remains to be regenerated to something like it was fifty years ago. Until then, or more likely just to save money, transporting animals will remain a do-it-yourself project fraught with hilarious fallout. Fallout literally, as when an animal falls out of a vehicle. Some brave souls even use their cars to haul smaller animals (how I once brought a lamb from southern Kentucky to northern Ohio) and learn that it is hard to get rid of the faint aroma of manure hanging over the back seats.

Until you have tried, in utter frustration, to carry or drag a one-hundred-pound pig physically onto a truck after all other methods have failed, you are not a true homesteader. If you have, in anger or desperation, used brute force to load any animal bigger than a little pig, I doubt that you are still among us, or if so, you have at least one hernia. (I have two.)

I have heard that a way to get a hog up a ramp is to put a bucket over its head and then back it up into the truck. If you believe that, then you believe Noah backed up two of every dinosaur species onto his ark by putting barrels over their heads. In fact, it might be easier to get a barrel on the head of a *Tyrannosaurus rex* than a bucket on a fear-crazed Hampshire hog.

A friend who has always been honest with me (so I believe him against all my experience) says he learned by accident that his dog was a born hog loader. As he struggled unsuccessfully to drive a hog up a ramp into his truck, he noticed that his dog was watching very keenly and closely. So he whistled and allowed the dog to go inside the pen. A nip here and a nip there, the way border collies will move sheep, and the hogs went right into the truck, my friend claims. They were more afraid of the dog than they were of the ramp.

I also know a clever guy who folded a panel of metal roofing around the pig he was loading to make a portable circular enclosure. The pig could not see out of its little temporary pen and so clever guy was able to walk it to the chute. This only works with a trailer that can be lowered nearly to the ground, allowing the animal to board without walking up a ramp.

The only reason many of us are still among the living is because of the invention of livestock trailers that can be lowered to almost ground level. Truck beds that are several feet above ground level look to be floating in outer space to a cow going up a ramp. She will not walk up it unless forced and forcing often requires the kind of actions that give the Humane Society, not to mention the farmer doing the forcing, heart failure. I bet that cattle ramps have killed or injured more cows and humans than all the foot and mouth disease outbreaks in history.

But just because you have the benefit of a livestock trailer that lowers to nearly ground level, thus avoiding an ascent to a truck bed, you are not home free. The beginner thinks he has but to back his trusty trailer up to the door of the barn so tightly that anything going out of the door must go into the trailer. Then all that is necessary is to "urge" the cattle or sheep or whatever up to the door and the animals will walk right on board. Would you place a bucket full of water right next to your farm pond and expect the fish to jump into it just because you urged them to do so?

You must make use of some kind of chute leading to the trailer door if you want to persuade the animals to walk on. Without a chute, they just scatter like birds disturbed at the feeder. Funneled into a gradually narrowing aisleway, the animals eventually realize that there is nowhere else to go but straight ahead into the trailer. That does not automatically mean that when one animal gets the idea and walks aboard, the others will follow. I have seen calves in a chute supposedly just wide enough for a single file, turn around and form another file headed away from the trailer. No matter what physicists tell you, two objects in a cattle chute can occupy the same space at the same time. Sheep have another tactic. Deciding to proceed no further when in the chute, they collapse in a heap and will not budge. On occasion, I have thought seriously of lowering a front-end loader on the tractor down to the prostrate ewe and lifting her onto the trailer.

You can buy very nice chutes, which also come in handy for worming and other handling chores. I never thought I could afford one, so I've just used gates made of boards or wire panels to form aisles, wider at the end away from the trailer door, narrowing gradually to the width of the door. Often you still must prod the animals along. Many a wise husbandman will park the trailer at the barn or pen door a day or two before loading and put choice hay or grain in it. The animals get used to the trailer and may even walk on of their own accord. We once parked a trailer out in a lot, the bed lowered to ground level, and put some yummy grain and molasses in it. After not getting fed all day, the steer walked right aboard.

In most rural areas, there are country butcher shops that for a nominal price (even if it is not nominal, it is well worth it) will come to your farm, slaughter your animals, and haul them to their butcher shop for further processing. In the case of a 1,500-pound steer, it is at least 1,500 times easier to load a dead carcass than a live animal. The slaughter guy will have a power hoist on his truck to lift the carcass aboard, and he will wrap it carefully so that it will stay reasonably clean on the way to his butcher shop. Some butchers bring a mobile butcher shop to your farm and do the whole meat-cutting job right there.

But even with expert help, be prepared for breathtaking moments. The first butcher we hired many years ago was a master of his work. He would walk up next to a steer with total calmness, dispatching it quickly and deftly with one shot from his .22 rifle. Unfortunately, some of his successors were not as practiced. Once, for no apparent reason, we had a steer go wild as if it sensed its execution. It broke out of the barn, and we had to hunt it down like a deer.

It makes one appreciate the drovers of yesteryear, driving cattle, hogs, and sheep to market over miles of roadless terrain. Cattle and sheep, then as now, were easier to drive than hogs. Robert West Howard, in his book *Two Billion Acre Farm*, writes about how drovers "knew that a hog with a wanderlust look in his eye and a nasty disposition calms down something terrific when it gets dark. . . . The drovers corralled the meanest boars and sows in pens, sat on them, and stitched their eyelids shut with hemp thread. It didn't hurt the hog much; after that he drove easier than a cow." If it were not for the livestock trailer, I wonder if resolute contrary farmers today might not resort to this method again.

As a boy, I helped drive our sheep several miles from our farm, when it was still owned by Grandfather, to graze on another of his farms. It was one of the most exciting jobs of the year. There wasn't much traffic then, and all the fields were fenced, so we just drove the sheep down the road. Neighbors, warned ahead of time, stood along their road frontage to keep the flock from wandering onto their lawns or barnyards. Usually there were no problems. The oldest ewes, remembering from previous years, would stick their heads up in the wind and march right along down the road to the new pasture. Occasionally, a farmer in our area still moves his cattle on foot from one farm to another, even in this day and age of heavier car traffic. It's a tradition here. In the 1800s, when this area was more cattle and sheep range than farmland, ranchers like Dave Harpster, after whom the nearby village of Harpster is named, drove herds and flocks all the way to Baltimore and Philadelphia. The cowboys of the western plains were preceded by the cowboys of the eastern forests.

Because the need for wheels and mobility infects everything in modern life, it is not at all surprising that the newest way to handle chickens is with what are called chicken tractors. Have you ever tried to drive chickens on foot? It's like trying to herd cats. As pasture farming grows in popularity, rotational grazing has become a practical way to raise chickens as well as cows on pasture, except when it comes to getting the feathered doggies to move along. The solution? Chickens will go into their coop when it gets dark. (Well, most of the time.) So put wheels on their coops and rotate the whole kit and caboodle to fresh pasture every few days. Old school buses make great chicken tractors. As agriculture returns to more small farms, we won't need so many school buses to haul children anyway because smaller, decentralized schools will increase, too. So we can turn old buses into chicken coops. Which reminds me of a most amusing story told to me by Paul Yoder, who farms contrarily in eastern Ohio. The buses he used for chicken coops still ran when he first converted them. Instead of pulling them from site to site with his tractor, as would later be the case, he would just drive them. Or his children would. Then he laughed and recalled, "There was nothing funnier to watch than the faces of children riding in buses going past the field on their way to school when they would look out and see a child driving a busload of chickens."

Chapter 8

The Cow Stable:
Health Spa of the Future

A sure sign that we are in some kind of new era in farming is that big-city newspapers report almost as much intriguing information about farming as farm magazines do. The Sunday *New York Times* of November 10, 2013, carried an article ("A Cure for the Allergy Epidemic" by Moises Velasquez-Manoff) about how we are suffering from an "epidemic" of allergies and that relief just might be as close as your nearest barn reeking with manure and murky with hay dust, especially if you are drinking lots of raw milk at the same time.

You think I'm joking. The latest study backing up this startling possibility was inspired by a curious observation: Amish farmers in northern Indiana, spending much of their lives tending livestock in their barns, were found to be remarkably free of allergies compared to urban populations. Mark Holbreich, an allergist in Indianapolis, investigated. About half of Americans have "evidence of allergic sensitization," but he found through testing that only 7 percent of Amish children on working farms were so sensitized.

This is just another example of how the soft economy of "get small and stay in" farming provides profit in ways the economists do not reckon. Having spent more of my lifetime in barns than in bathrooms I represent some evidence of this theory. The only thing I know for sure that I'm allergic to is TV reality shows. By spending so much time stomping around in manurey cow barns and dusty haymows, and drinking lots of raw milk (easily a gallon a day in my twenties), I gained a life free of allergic distress, if this new study is correct. I have long considered my barn to be my church, and now it turns out to be my health spa, too. All for free.

The traditional Amish way of farming on a small scale provides a model for the new garden farm. It shuns the "get big or get out" model and still manages to be quite profitable while flying blithely along under the radar of the industrial economy. The typical Amish farm, in fact, is a most amazing example of true home economy. Very little needs to be purchased off-farm, so overhead costs are low. Most of the food for the farm family and for the livestock is raised, not bought. The farmer's house is part of the farm and, generally speaking, sells with the farm, its considerable cost not reckoned as a separate expense. The farm family supplies most of the labor and does not have a payroll it must meet every week. Traveling by bicycle and buggy avoids the enormous cost of automobiles. Using horse-drawn machinery avoids the enormous cost of tractors. (The latest John Deere harvester with two 40-foot headers costs a half million dollars, and there are others on the drawing board that will cost a million.) In fact, many Amish farmers actually make a nice profit selling extra horses to wealthy draft horse enthusiasts who raise the animals for a hobby. The Amish farm's livestock manure provides the fertilizer for cropland, saving something like $600 per acre in purchased fertilizers. The most blessed advantage of their farming is that they can walk to work across the barnyard and walk home again the same way without cost or commuting time. When they do "commute," they often ride bicycles. They eat mostly at home. Most of their farms have a second house for retired parents, avoiding the cost of retirement homes. Most of their clothes are homemade. Many of them have gas utilities in the house and diesel-generated electricity for milking the cows. In many cases, windmills on hills above the house fill their cisterns, and the water flows by gravity into their homes. It takes longer to draw water for a bath and longer for the toilet to fill after a flush, but who cares? The Amish are also pioneers in the use of solar-generated electricity. How can anyone call them "backward," especially now, when very up-to-the-minute executives are bicycling to work and rich people ride in horse-drawn carriages for recreation?

For the sake of honest journalism, I have to mention my wife's brother, who spent most of his working life in a dairy barn but who occasionally suffered severe asthma attacks all his life. I also know of farm boys so allergic to hay that they had to quit livestock farming. At least that's the excuse they gave. Maybe farmers seem more immune to allergies because over time their occupation weeds out the ones physically not fit for it.

But I like the idea that barns and raw milk have been my allergy salvation. Strangely enough, the scientists leading the way in this investigation do not advocate drinking raw milk. It can contain deadly pathogens, they note. Note that they say this almost in the same breath they draw to point out the healthfulness of Amish children, who drink only raw milk. Nor do the scientists advocate more small livestock ventures, as the unrevolution surely does. They want to find a way to isolate the rich microbial life in the dirt of the livestock barn and chicken coop and apply it directly to allergy victims who live high in their sterile city apartments. So now we have one more product that the local food and farm movement can sell: bottled microbial life from our barns along with bottled raw milk. We need a brand name here. How about "Barn Aire?" A sniff a day keeps the asthma away.

Barns have other social advantages, even more modern ones. For instance, if during courtship you want to find out whether you and your beloved will get along okay in marriage, spend some time milking cows together, like in the operator's pit of a herringbone milking parlor. That's what Carol and I did, occasionally dodging the rich microbial life spattering down on us from the cows looming above, compared to which allergies and asthma seem preferable. I figured if she could endure a week of milking cows in a "parlor" (however it got that name is another one of those unsolved mysteries), we could probably endure marriage for at least a century. Maybe if we continue to spend a lot of our time in our health spa barn and chicken coop, we will last that long, too.

Chapter 9

~~~

The Rise of the Modern Plowgirl

S ociety is in the habit of using the masculine pronoun when referring to farmers, another indication of the lingering, perhaps unconscious, prejudice toward women. Or for those who uphold the inviolate female as being above such nasty, lowly tasks as mucking out the hog pen, the attitude could be interpreted as a lingering prejudice against farming. But no matter, the most obvious and promising sign of the new agriculture is the leadership that women are taking in the unrevolution. Women have always played the key role in farming, of course, but they have seldom been given credit for it publicly or historically. Farming is a man's world, American culture wants us to believe, and as is true of all culturally treasured myths, no amount of plain everyday evidence to the contrary matters much. In many cultures, in fact, women do most of the farm work, and in traditional American farming of the recent past, the farms most likely to succeed have been those where the wife worked right along with her husband. (The most unheralded heroines of agriculture have been those wives who would much rather be farming but who take a job off the farm to get the insurance and extra money, without which their husbands could not afford to farm.)

Farming is most successful where husband and wife cooperate in all the work, including the housework and cooking, but in the past that was not the usual case. The "missus" was supposed to do not only all the cooking and housecleaning but do it in her spare time when not caring for chickens, growing a garden, and milking cows. Old male farmers even today look down their noses at chickens and gardens as "wimmenswork" and forcing men to cook a meal they view as a kind of emasculation. Real farming is riding around on a tractor with its motor throbbing away between their legs, even when that threatens to cause something really approaching emasculation. It all started down that road in the hunting and gathering age, I suppose. Men did the hunting (adventure) and women did the gathering (boring) because it could be done while caring for children. Then when settled farming came along, the men naturally gravitated to the heavier fieldwork like plowing lest they have to help dry the dishes. They pointed out that it was almost impossible to lift a horse plow out of the furrow and reset it in the next furrow while carrying a baby in a sling, never once considering the possibility that just maybe they could stay home some of the time and take care of the babies themselves. When slaves were no longer available to pick cotton, share-cropping women more than men stepped in to do the job. Joe Dan Boyd, my contrary friend and coworker at *Farm Journal*, fondly recalls as a little boy riding on the cotton sack his aunt pulled between the rows as she picked.

At any rate, after the plow became the symbol of agriculture in America, the role of women in farming receded from the public eye. Women were supposed to stick to the kitchen and garden and leave the real business of farming to the "menfolks." When I interviewed farmers and their wives about their business, it was amazing how often the wives answered my questions much more readily than their husbands and how they so often did this by diplomatically and cleverly putting words in their husbands' mouths. It is obvious that behind many successful farms lurks a wife smarter and more articulate than her husband, and sometimes the husband was wise enough to know it. When he was not, the wife knew how to keep the male crest from falling by seeming to defer to her husband on every occasion. She might begin a sentence with "as my husband well knows" and then make an observation that husband never thought of before. The wives knew they had to make their mates look like top operators so that they could borrow the money they needed to keep on going. Bankers prefer to deal with men. They subconsciously did not think women were smart enough to run a business like farming. Or perhaps they knew the women are smart enough to quit borrowing so much money.

The prejudice was prevalent even in farm magazines. *Farm Journal* included a "Farmer's Wife" section in the back of the magazine filled with recipes and growing flowers and folksy charm about farm life. The real hardcore business of farming went in the front of the magazine. Amazingly, no one seemed to see any prejudice on display. I asked one time, during my stint at *Farm Journal*, what would happen if we put a section in the back of the magazine designated as "The Farmer's Husband." The women editors laughed; the men did not.

The best evidence of this prejudice that I observed during my tenure as an agricultural journalist occurred when *Successful Farming*, our rival magazine, decided to drop its "woman's section" altogether, under the notion that "women's work" had no place in the real he-man world of modern, large-scale farming. What did women know or care about 300-horsepower tractors? The decision cost the magazine a bundle of money because readership dropped precipitously. As it turned out, farm wives knew quite a bit about 300-hp tractors, especially about whether their husbands could afford them or not. It also turned out that the wives read the whole magazine more closely than the husbands read any part of it. When circulation started to drop off, the editors reinstated its woman's section and asserted how important women were to the "decision-making" process on the farm.

Eventually, however, Monster Farming did make farm wives more or less disappear. Of course, it made farm husbands disappear, too. Instead of a family farming together, a top operator (*Farm Journal* actually started a magazine by that name) ran his operation with a hired staff of workers and spent most of his time keeping landlords happy. (One such top operator I wrote about in the magazine of that name had fourteen landlords at one time.)

But now with local garden farming on the upswing, the plowgirl is making her presence felt. The times are right. Often her husband has a job off the farm and she has no choice but to take over when he is gone. Just as in primitive times when the husband was off hunting, she becomes the farmer while her husband is off working on an assembly line or in an office. She doesn't have to worry about heavy lifting because it is all done now by the touch of a button actuating a hydraulic lift. Starting her own farm and food business is not a taboo anymore, and in marketing local, artisanal foods she has a good chance for success. She understands better than most men the vital connection between food on the table and food in the field. The *Edible Cleveland* magazine (edited mostly by women), has an article ("Ohio's Cheese Women" by Jean Mackenzie, Spring 2013, page 30) about female farmers who turn the milk they produce from their sheep, cows, and goats into cheese. A photo shows eleven of them sitting on hay bales in poses that show clearly they are up to the challenge of pitching manure, making artisan cheeses, and caring for their children at the same time. Then on page 24, there's an advertisement for Cleveland Independents, a group of more than 80 locally owned and operated restaurants, featuring young male chefs—thirteen of them pictured—who are shaking up the Cleveland social scene with their independent restaurants committed to local food and fine dining. Although the comparison is not exact, it is amusing to think about how women were once society's cooks and men did most of the farming, while today the situation often is the reverse.

There's another way that women are becoming the new stars of farming. They are, generally speaking, more talented at interacting with the public. There are always exceptions of course, but men who are comfortable with farming generally become very nervous if thrust into the public eye. One reason they farm is because they like solitude, at least most of the time. This is true with farm women, too, of course, but it's not nearly so pronounced. Many a male farmer won't even answer the phone if his wife is in the house. Then he stands there, behind her, and tries to coach her about what to say over the phone. Plowgirls become, by default if for no other reason, not only producers of food, but retail sellers, too, which is just as important in the garden farming unrevolution as the actual fieldwork.

If you can't use a computer these days, it is almost impossible to run any business, and in my experience women are better at transitioning over to computer clicking. I would be lost without Carol in this regard. She can understand computer logic so much better than I can. I have a theory about it. The best training for learning how to deal with computers is to raise babies. If you become adept at the simple yes–no way of thinking that works fairly well when dealing with children, you can think your way through a computer problem more adeptly.

Finally, women have, in my experience, more patience than men, and patience is more vital in farming than it ever was, especially when dealing with government regulators. With more and more backyard food businesses springing up, more and more government regulation springs up, too, and some of it can be maddeningly irrelevant. For example, no matter how much evidence piles up about the inherent safety of raw milk in modern times, the government, under the influence of the pasteurized milk industry (which sees raw milk as a threat to its bottom line, or sometimes simply is ignorant of how milk is produced on farms today), continues to enforce outmoded regulations. At least that is my opinion, having consumed only raw milk for many years. Women, for centuries the subjects of male domination, know how to handle the pasteurized milk regulators better.

Of the newer plowgirls I know, Anna Wills strikes me as typical, although there is nothing typical about any of them. Her husband, Brent, came to visit us and that's how I know about her. As he regaled us with stories of their Bramble Hollow Farm in Virginia, he mentioned, almost casually, as if he were saying nothing unusual, that Anna had caught a black snake eating their turkey chicks. I stopped him right there.

"Please? Did you say she caught a black snake with her bare hands?" I'm sure my eyes were bugging outward.

"Well, no, not barehanded." He still seemed very casual. "She had gloves on and caught it up in a sack."

"She caught a black snake big enough to eat young turkeys?" I was still incredulous.

"Well, yeah, we dissected it. Had two poults in it."

I'd heard of stories like this before. In some cases, the animals found inside the black snakes were still alive. But prim, well-educated, sophisticated members of the so-called "gentler sex" catching them? This is the twentieth century on the farm, not the nineteenth. When I interviewed Anna by email, that's the first question I asked. She replied by sending me a photo Brent had taken of her holding the snake. Awesome. It was almost as long as she was tall. I have had my share of confrontations with snakes but never captured one even half the size of this one. She said that she managed to put the snake in a sack and waited for Brent to come home and dispatch it. Other photos showed the remains of the turkeys that it had swallowed but that were still undigested. "Actually, if they behave themselves, I like having black snakes around," Anna pointed out. "They eat rats and mice."

How does a woman with a degree in Environmental Science, who had worked previously as chief ranger for the Virginia state parks, become a farmer with two young children, homeschooling the older one now, and doing much of the farm's marketing, bookkeeping, scheduling, and correspondence, plus helping with the chores? "I never imagined this would happen to me," she says. "Now my entire life revolves around food and feeding people and animals."

"I am a mother to my children and a wife to my husband first and foremost," she continues. "That may sound like something out of the fifties, but that's the way it is." She more than once pointed out that Shannon Hayes is an important role model for her. Ms. Hayes is well known in the local food movement for her biographical book *Radical Homemakers*, which details her life as a committed mother and farmer. Anna calls herself a radical homemaker, too.

If you want to be impressed, visit BrambleHollowFarm.com. Anna and Brent raise all kinds of heirloom livestock and poultry, sell garden produce at farmers' markets in the vicinity, supply a Roanoke restaurant, and sell their meat through their own CSA. Somehow they also find time to make and sell wild raspberry wine and various artisanal breads. If that is not superhuman enough, Brent also works off the farm as a certified soils consultant.

Somewhat in contrast to the Willses, champion market gardeners Jan Dawson and her husband Andy Reinhart (mentioned first in chapter 1, "No Such Thing as '*The* American Farmer'") have no children. They are not particularly fervent about what we call old-fashioned family traditions except for a loyal and loving relationship with each other and their siblings. They are vegetarians. Both are very oriented toward natural diets and natural medicine. Andy abstains entirely from alcohol. Jan might have a bit of whiskey when she visits sinners like Carol and me. With his ponytail and beard, Andy looks like an old hippie. Jan dresses somewhat like the rural women of yesteryear with whom she mostly disagrees philosophically. Andy finds Zen Buddhism interesting, but only to a point. Both side with liberal views in religion and politics. But they think that spiritual values are essential to a wise life. Both of them are serious, dedicated organic farmers, so serious that they think some of the certification standards are hypocritical. As a result, they have not sought certification for their produce. "How can a food be labeled organic if it is shipped in from hundreds of miles away?" Andy often asks.

While Jan generally voices what I call classic liberalism in politics, she is conservative in her personal economic philosophy, believing that indebtedness means loss of personal freedom. Time and again I have seen her, and Andy, decide against expansion of their farmers' market business simply because they didn't want to work any harder and didn't want to borrow money. Much of what I say in chapter 2, "Farming Is All About Money, Even When It Isn't," about contrary economics is inspired by them. They want time to study and enjoy the natural life and beauty they find on their little, mostly wooded farm. Better to reject consumer-driven materialism, they say. Better to live a little on the poor side.

The contrasts between various plowgirls (and plowboys for that matter) are what make them so intriguing. When I meet new ones, I try to be coy about what I say until I learn their views on politics and religion. I seldom find out. As the old saying goes about contrary Maine residents, they often vote one way and drink another. They won't predict what they will do the next time, either. They may support a very particular religious denomination or they may not support any of them. They strongly dislike being labeled. And they belie most labels. When I pressed one of them on religious belief, she finally sighed and said in a slightly exasperated tone, "I just don't *care* about institutional religion. Whatever you want to believe is okay by me. I'm just not interested in the subject." That is a remark I hear more and more frequently from millennials of either gender, in all walks of life. Could the heyday of institutionalized theology be passing along with the heyday of institutionalized farming?

CHAPTER 10

Finding and Keeping a New Age Farm Partner

There are hardly any garden farmers so contrary that they can go it alone and last long enough to brag about it. Just from the standpoint of physical participation, farm work often requires four hands rather than two, and mental decisions are almost always better coming from two heads rather than one. (Three or more heads can make it worse.)

But the bigger reason that farming partnerships, especially the love and marriage kind, have a better chance for success is that farming is a lonely occupation. Most farmers like solitude in the sense that they are uneasy in crowds, but loneliness is no fun at all. The first lesson of this biological fact of life is that if you are a mad farmer as Wendell Berry describes in his poems, you need a mate or partner who shares your madness. Many farming ventures collapse because only the husband or the wife really likes living that way. Out of love, the other agrees to go along with the idea but eventually learns that he or she can't stand uncomfortable and seemingly unrewarding physical work twelve or more hours a day doing something that is not at least partly his or her idea. His or her lack of interest ends in divorce or in a kind of sullen silence that is worse than separation.

It is not difficult to find contrarians if you are one of them. If you put two contrary farmer types in a convention hall crowded with people, it will take only a couple of hours for them to find each other. That's how I first met writing farmer Mike Perry. We just hit it off before we knew much about each other. I think maybe contrary farmers give off some strange magnetism that other contrary farmers feel. Maybe they give off an odor that other contrarians can smell, like the way a raccoon can find the nicest, ripest ear of corn in a ten-acre field in less than an hour.

Getting along with a contrary farmer in marriage involves lots of reverse psychology. Let us say you, as the wife, would like to have a patch of wild black raspberries in the garden so you don't have to thrash around in the woods to pick some. So you opine, where hubby can hear you but not speaking directly to him, that it would be really dumb to move some plants from the woods to the garden. No one has probably ever done that before, you add. Probably wouldn't grow there anyway. If you are married to a contrary but loving farmer, you will be picking wild black raspberries from the garden in a year or so.

Hard physical work always seems easier with two people involved: "Many hands make light the task," as the old saying puts it. When I am hoeing corn, it makes all the difference in the world when Carol's hoeing toward me from the other end of the row. Instead of inching along growling about the damned flies and gnats, the damned hoe that needs sharpening, the damned sweat in my eyes, and the damned weeds that would have been easier to hoe out last week, I'm thinking, "Oh, great, I only have to do half the row."

The other secret to happiness in spite of the unpleasantness of some jobs is to extend togetherness to all areas of work. Males, especially older ones, tend to expect "the missus" to help out with fieldwork, but husband is nowhere in sight when it comes to doing the laundry or the housecleaning. The Hubbards, mentioned in chapter 1, "No Such Thing as 'The American Farmer'," did everything the old manual way because they understood there was more economic freedom to be gained. That meant that Harlan helped Anna with the laundry, the housework, and laying by their food.

Shelling peas alone is boring, to say the least. Two people shelling and gabbling to each other or even just silently aware of the other's presence, makes it a rather pleasant activity in an odd sort of way. At least you can do it sitting down in the shade. Or as Elsie Kline writes in her delightful column, "The Farm Home" in *Farming* magazine (Fall 2014 issue, page 46): "Growing up, and also with our own children now, we'd sit in a circle, capping strawberries, shelling peas, shucking and cleaning corn, snapping beans, peeling peaches and pears, singing and playing word games as we worked. What fun it was!"

Of course, if both partners lean toward contrariness, working together takes some getting used to. I am fast and sloppy; Carol is slow and neat. The way I stack wood drives her crazy, and invariably my stacks start to fall over in the second year of drying. The way she does it makes me think we will freeze to death before the wood gets stacked at all, but with her way it will stay stacked until eternity if not used before then. Solution: she stacks and I split. Even as slow as she is, she can keep up with my splitting.

To find a compatible mate these days, farmers are hitching up their computers and trotting off down the internet highway. Sometimes this method of courting is actually successful. More often it is not. The main reason why Joe or Mary can't find a mate in their own neighborhood or community is the same reason they can't find one electronically either. They are *too* contrary.

I cheer when lonely Joe, who is busy milking cows way out in Deserted, Iowa, finds love and marriage with lonely Mary in Last Chance, Illinois, by way of computers. But I think it is sort of sad that there are lovelorn websites now tailored to appeal to "farmers only," as if a tiller of the soil can't often find a mate in town who would love to live on a farm. Reminds me of growing up in a world where Catholics were reprimanded for dating Lutherans, and vice versa. I can name any number of cases where ignoring this precept and joining in a "mixed marriage," as our clergy called it in pitying tones, was the best thing that could have happened to both husband and wife. They learned to view religion with more objectivity. Similarly, a whole lot of farmers can thank their lucky stars that they married someone who might not be able to tell a soybean from a corn kernel but who could hold down a job in town that paid the bills until the soybeans and corn became profitable.

Online dating will soon be as acceptable as going to church socials to find prospective mates used to be. If you really believe that love and marriage go together like horse and carriage, you know that choosing a life partner is your most important decision in life. The internet can widen the playing field, as long as the players are aware of all the pitfalls involved. It certainly is no more risky than trying to find a mate in a bar, and nowadays that's also a fairly common occurrence—and has as good a chance of turning out as successfully as finding a mate in church.

What bothers me about search for a farm partner on the internet, the pitches that websites make to advertise themselves. Some of the ads project grossly prejudicial and stereotypical views of the farmer—a simple-minded, "aw, shucks" character that I thought had vanished from the social scene years ago. If any other group—social, occupational, religious, or ethnic—were represented with such crass inaccuracy, there would be outcries heard around the world. But farmers, aw, shucks, they are used to that kind of prejudice. It is just so far-out wrong, we just, aw, shucks, shrug it off. And that kind of image-making unfortunately works both ways. In retaliation, some rural people respond by ridiculing "city slickers" just as ignorantly.

But the whole notion of hunting for a mate online is problematic. Am I supposed to believe that farmers are too cut off from mainstream society to meet possible mates and therefore need help? Or are they too busy farming to take part in social activities where they can meet people looking for love and marriage? Come on now. Is there almost anyone in this day and age so cut off from the mainstream that they can't find someone to date? I don't believe it, unless maybe if you live in the middle of the Sahara Desert, and even then I'd bet against it. I spent my wild youth in a seminary located out in the woods, studying for the celibate priesthood. You can't get any more isolated than that, but believe me, there were still plenty of girls around if one had a mind in that direction. Sex has hardly ever been stymied by anything so easily out-maneuvered as isolation. If you can't find a mate, your problem runs a whole lot deeper than where you live or what you do for a living. When I left the seminary, I went back to farming, milking a hundred cows in a very rural area. That's hardly mainstream. But it often seemed to me there were girls peeking out from behind every crossroads stop sign in the county. And one is never so busy as not to find time to peek back. I know a divorced farm guy who has gone the internet route to courting, and between text-messaging and chatting on his cell phone and traveling around checking out prospective mates, he racks up more hours at home than he used to complain about when he was married.

When two contrary farmers from far apart do fall in love, the outcome is doubtful because one of them will have to move. If you marry someone from far away, you will spend a lot of time traveling and visiting your in-law family. This can become a real burden when a farm is involved. If you get past that hurdle, there comes a time when help is needed in caring for children or parents, or grandchildren and grandparents. Lucky above all are the parents whose children either remain close to home or return to the home grounds to live close by. If you are a farmer with your heart buried in your land, it is important to talk this over with a prospective mate. I was lucky again. When Carol and I decided to move back to farm country, we could have moved to her old stomping grounds in Kentucky, but she was kind enough to agree to settle on mine.

We have been married for over fifty years now, and I know for sure that the success of our marriage wasn't because of the number of people we dated and analyzed beforehand to see if they met our lofty standards of perfection. Neither of us had done much dating before at all. I was so madly in love with her that I was unable to make any kind of rational, objective judgment about anything. I am asked sometimes what is the secret of a successful marriage. I haven't the slightest idea, least of all anything a computer might have told me.

In the rural America where I live, the idea that a farmer can't find the time or place to meet prospective marriage partners is preposterous, or at least quaint. The few marriageable young farmers here in my neighborhood not already spoken for stand to inherit so much wealth that every guy or gal in town who knows the score dreams of snagging one of them. Farmers don't have to go hunting far and wide for mates. They just have to sit out in their lonely cornfields in their super-tractor cabs and wait for knowing suitors to show up. The internet courter mentioned earlier tells me that when a gal sees a guy standing in front of a half-million-dollar tractor outfit on land selling for $8,000 an acre, she smells money. The farmer's problem, male or female, is the same one it has always been for the richer among us. Does he or she like me or my money?

But the whole notion that farmers live apart from society and therefore may have a hard time finding a mate is outdated anyway. Garden farming is blossoming in urban areas even more than in rural communities, and both genders are heavily involved.

CHAPTER 11

Big Data and Robot Farming

The news these days is all about how cars may someday drive themselves and how tractors already can. Am I supposed to be impressed? A hundred years ago, everyone had a self-driving car. It was called a horse and buggy. If you fell asleep at the reins, you could depend on a good buggy horse getting you home safely. Arthur Hertzler, in his often amusing book, *The Horse and Buggy Doctor*, regales readers with stories about falling asleep in his buggy as he drove for hours to get to the bedsides of ailing farmers far out in the country or as he returned home. His faithful horse would take over, not always with the best results but not ending in tragedy, either.

In farming, we have had "tractors" that could drive themselves for centuries. They are called draft horses. Like smart phones, they respond to voice commands while the farmer walks along beside them, filling the wagon they pull with corn or grain shocks or hay bales or whatever. Once, after leading our old Flora many times back and forth across the barnyard as she pulled grapple fork loads of loose hay by rope and pulley from the wagon in the barn up to the hayloft, I decided she could probably make the trip by herself while I sat in the shade. And so she did, but decided to take a bit of a detour over to the horse trough on her way across the barnyard.

Dad, handling the grapple fork, did not think that was as funny as I did.

My uncle Lawrence Rall may or may not have been the first farmer to actually figure out how to make a real tractor drive itself. He noticed that when his little Ford 8N (circa 1947) was pulling a two-bottom plow, one front wheel and one back wheel in the furrow, it tended to stay on track without much guiding until he got to the end of the field. Too bad fields had to have borders. It got him to thinking. Riddle me a riddle. What kind of field has no end? A round one, of course. It occurred to him that if he plowed in a circle rather than straight, the pull of the plow would keep the tractor tires in the furrow and he would not have to guide at all. So he gave it a try. He made a circle with the plow out in the center of the field and, after a few rounds, sure enough, he could take his hands off the steering wheel altogether. He finally worked up enough nerve to sit in his pickup while the tractor and plow went driverless for a couple of rounds. I saw it with my own eyes. I was only a child then, but I never forgot. Of course the fence corners and the middle of the circular headland had to be plowed later, so round-field plowing wasn't all that efficient. Grandfather Rall, who was still more or less running things, did not think Lawrence's feat was one bit amusing and put a stop to robotic tractors for the time being.

But my other grandfather used the idea to invent a driverless lawn mower. He set a stake in the middle of his large, grassed barnyard, tying one end of a rope to it and the other end to the lawn mower on the outer edge of the lawn. Sure enough, the mower went round and round as the rope wound around the center post. Other Grandfather sat in the shade and cackled.

He also took notice of uncle Lawrence's feat and devised his own version of a driverless tractor to shorten the time it took to plow and disk his fields. My uncle Maury swears this story is true, and if you knew Maury like I do, you know he never lies. At that time, Grandfather's tractor was a lumbering old Fordson that also would stay in the furrow as it plowed, but its speed left a lot to be desired. Grandfather realized that he could disk the plowed areas with his team of horses faster than the tractor could plow it. So after he had plowed a couple of rounds with the tractor, he brought the team to the field and hitched them to the disk. After he got the tractor started across the field with the plow again, he raced over to the horses and disked across the field, passing up the tractor and plow in time to turn around at the other end of the field, race over to the tractor in time to turn it around before it lumbered into the fence, start it back across the field again, then race back to the horses and disk, passing up tractor and plow again, and repeat the turnarounds at the other end of the field. The only thing that kept the operation from being a success was Grandfather's stamina. After several acres, he just couldn't keep up the pace anymore.

Dad discovered yet another way to convert a tractor into driverless mode. The one-lane paved road past our house was ridged ever so slightly in the center. For some reason I've never figured out, when we drove the Massey Harris Challenger on it, the front tires of this tricycle model would stay on that ridge. If the tractor started edging ever so slightly to the left, the right front wheel would pull it back to the ridge again. And vice versa. There was a gentle *S* curve in the road between us and our neighbor, and sure enough, the tractor wheels would follow that curve faithfully. We never got up enough nerve to let the tractor visit our neighbor alone however. He was quite old, and we were afraid he might have a heart attack if he saw a self-driving tractor headed his way. It would be sort of like seeing Washington Irving's headless horseman galloping down the road.

Today's robotic tractors can't be trusted alone either, although that is being remedied. They will drive themselves across the field okay, but someone has to turn them around at the appropriate times, like when a country road is looming up ahead. A great story is told in our area about this, and I know the people involved and the terrain where the incident supposedly took place, so I believe it. The tractor "operator" with nothing to do as the tractor and disk journeyed across the vast moonscape of a modern field, fell asleep. His monster robot rumbled along right smartly and halfway across the gravel country road at the end of the field before he woke up.

A few years ago, I thought that the most sensational news in robot farming was the introduction of milking stalls where the cows essentially milk themselves. No one needed to put the suction cups of the milking machine on the cows' teats or take them off. Of course these amazing robots cost so much that the farmers who decided to keep on milking the old way, and invest the savings in the right stocks or in more land, might make enough money to stop milking cows altogether. Such farmers could retire and rent robots to go fishing for them.

Actually, if you think about it in a slightly joking way, money is itself robotic. You can "put it to work for you" in bank derivatives and sit around doing nothing more than counting your profits. You can let your robotic computer do the counting, too. Of course, you and your computer might spend about as much time subtracting as adding.

But now something even more robotically amazing seems to be in the works. International agribusiness companies think they can invest in enough of what is now called "Big Data" to not only drive machines but garner all the information about agronomy, weather, and climate change needed to know what crop to plant where on which day, all while never leaving the office. Then big companies can get into the crop insurance game along with the government, as some already have. The theory is that they will be so smart with all that computer wisdom that crop losses will be kept at a minimum hitherto unknown to agriculture, and then big insurance payouts won't be necessary, either. Everyone would win, don't you see? A spokesman for Climate Corporation, a crop insurance company that Monsanto bought in 2013 (and has already sold, I hear), gave as an example of what Big Data had already revealed to them that corn grown in some parts of Kansas doesn't really pay—as if we haven't known that as far back as the Dust Bowl days of the Dirty Thirties. According to an article in *New Yorker* magazine ("Climate by Numbers" by Michael Specter, November 11, 2013), Climate Corporation's scientists could "process fifty terabytes of weather information every day, roughly the equivalent of a hundred thousand movies or ten million songs. The data include eight years' worth of soil, moisture, and precipitation records for each of the twenty-nine million farm fields in the U.S."

Did you know there were twenty-nine million farm fields in the United States?

The article goes on to say that the company could create "moisture and precipitation maps so precise that in some cases using computers, a farmer can determine whether the field on one side of a road is wetter than the field on the other side." Going out to the field to check growing conditions, one of the most enjoyable parts of farming in my view, will no longer be necessary. Neither will farmers be necessary. Since the tractors and equipment can also be operated robotically from a computer twenty-nine million fields away, and the computers know what should be planted where and how, who needs farmers? Why, shucks, a savvy farmer could live in a luxury high rise in New York City and manage his land all over the world. Droughts and hailstorms and all that could be tracked and analyzed to provide information about where and when good weather will ensure profitability, or at least not too much loss. High-risers could peruse minutely detailed, computer-collated statistical records on every raindrop, every temperature degree, every whisper of wind, every oscillation of every ocean ripple, every zig and zag of every jet stream, every wiggle and waggle of every polar vortex. The computers can sense and record soil moisture so minutely and in such detail that they can record which one of a farmer's cornfields he used to pee in yesterday. The data can show, unerringly, when and where to plant wheat in Russia, soybeans in Brazil, and corn in Sleepy Eye, Minnesota. No more guesswork, no more risk. Data will rule. If you really think you must see what's going on in one's far-flung empire, send a drone. Hear the cash register of overhead costs ka-chinging all across the land.

Stay-at-home farmers don't need drones. In fact, their favorite pastime is walking over the fields, watching how the crops are growing. In dealing with uncertain weather, they rely on time-tested traditions and local history, which are free. If you look at agricultural history the traditional way, it becomes apparent that many of the calamities in farming that computer models now blame on bad weather or climate change are really caused by human behavior. If the dust storms of the 1930s occurred today, climate change would no doubt be blamed, whereas the problem really came from too many farmers thinking they were going to get rich plowing up the prairies for wheat (as they have been lately plowing them up for corn). Today's bad drought in the West might be linked to climate change, but water shortages are as much the result of increased populations as the ebbs and tides of El Niño and La Niña. Interestingly, in the July, 2014, issue of *Acres U.S.A.*, the legendary organic farm advisor, Amigo Bob Cantisano in California said about the same thing. After pointing out traditional ways of coping with drought in what is essentially a desert area, like the way some wineries have learned to grow wine grapes that require little or no irrigation, he said, "My friends who are practicing dry farming are not going to notice there is a problem, or very little. . . ." If this trend continues, eventually Monsanto, Cargill, Dow, and the rest can cover all twenty-nine million fields with plastic domes like those over sports stadiums, a giant-sized version of what I call in chapter 3 "enclosure farming." Big Ag will have accomplished its fondest dream: a monopoly of food production from seed to shelf, from plant to plate. It can patent all the plants that grow under the domes and patent the weather there, too.

Small-scale robotic farming will also be possible. Robots can perform many tasks hitherto thought to be the role of humans only. They are being designed especially to work in hazardous situations such as rescuing real people from burning buildings.

By the same token, we should be able to program robots to plant gardens, pick strawberries, hoe weeds, dig potatoes, chase away deer and raccoons, pull bindweed, and brag about corn yields. This may become the best way to control weeds that are immune to herbicides. Armies of robots, programmed properly, will be able to distinguish pigweed from soybeans and roam the fields hoeing out the former.

The eerie insanity of a robotic world will not end there. An article in *Farm and Dairy* ("IBM Unveils Five Innovations That Will Change Our Lives," January 9, 2014) observes that "everything will learn—driven by a new era of cognitive systems where machines will learn, reason and engage with us in a more natural and personalized way." There are 3–D computerized printers now that serve up food even as delicate as chocolate drops. Cooking stoves are well on their way to complete robot status. Someone will still have to put the food to be cooked in them (maybe), but the stove will do the rest all by itself. There is even computer talk of a program that would allow a person to go on tweeting even after he or she is dead. Far-seeing scientists talk about being able to computerize an individual human's genetic makeup—mind, emotions, and character—so exactly that a robot with your DNA inserted into its robot genes would, for all practical purposes, be you, and you can go on living indefinitely in a "roboticized" state of existence.

What will humans do in the robotic era? The standard answer is that there will be plenty of jobs making and maintaining the robots. But that is unlikely because robots are taking over assembly line production already and have the ability to correct their own errors and make repairs. The robot car will be built and maintained by robots that will be overseen by correction robots. The robot car will have a built-in drone that can flit around and send you back photos of all those places your car can't go. You can relax at home and impress your friends with the great photos your drone takes.

But, says the robot industry, humans will still have plenty to do just "monitoring" all the robotic machines. Someone will have to make sure the robots keep doing what they are programmed to do. Aha. Since in farming, if not in everything else, humans are not in agreement about what needs to be done, we can speculate that very soon organic farmers will program robots to hoe out genetically modified crops, while chemical farmers will program robots to spray herbicides on organic crops. The long-predicted future where robots fight wars will move into the farm fields.

Overpopulation of robots will surely become a big subject of debate, especially among the robots. Since they will be programmed to think in a completely logical way, they will decide to get rid of humans. Who needs them anymore? More trouble than they are worth. Don't need agriculture without humans around anyway. Robots don't eat food. Can't get farming any more efficient than that.

CHAPTER 12

The Invasion
of the Paranoids

I f you have not yet been invaded by aliens, brace yourself, because you soon will be. There are so many invasive plants and animals and humans approaching from all directions that there is no escape. It is not proper for me to make fun of something that is not funny, but since I have been invaded, too, maybe I can be forgiven.

Currently, my favorite danger-of-the-day is the Invasion of the Tumbleweeds. It happened in Colorado, and to the ranchers there it's not a bit funny. I quote from an Associated Press story ("Colorado Tumbleweeds Overrun Drought Areas" by P. Solomon Banda, April 8, 2014): "Mini-storms of tumbleweed have invaded the drought-stricken prairie of southern Colorado, blocking rural roads and irrigation canals. . . ." I now sing one of my favorite songs with my fingers crossed: "Drifting along with the tum-ble-lin' tum-ble-weeds. Cares of the past are behind, nowhere to go but I'll find just where the trail will wi-ind. . . ." Cares of the past are behind? No more. Today, the trail always winds back to more trouble.

If you have not been invaded by tumbleweeds, maybe you are in the path of the feral hog invasion. This, too, is not at all funny, even though, far from the hog wars myself so far, I can't suppress a little giggle now and then. Hogs that have gone wild are costing us $1.5 billion a year, says another AP story, including $800 million to farms in some thirty-nine states. I smile because I am sure that the US Army could solve this problem in about a month with a few boots on the ground. The real problem is that society as a whole doesn't think it is a problem yet. When the wild hogs invade Main Street, then the solution will come overnight. Lots of pork barbecues will follow. I bet I'm not the only one who hopes that the disease that is killing thousands of domestic pigs, porcine epidemic diarrhea (another invasion), moves into the wild population. But I doubt very much that will happen because wild pigs don't have to put up with a life of confinement. If you raise a few hogs the old-fashioned way, allowing them room to move around, hopefully on deep bedding that stays fairly dry on top and somewhat warm from the composting process in progress underneath, I doubt you have anything to worry about.

The climate change invasion is the biggest cause of social paranoia at the moment. Obviously, we should be concerned. Actually we should be cutting down on the amount of carbon we are expelling into the air, whether the climate is changing or not. But despite all the hand-wringing, I have not yet heard of one serious governmental proposal to cut down significantly on the amount of unnecessary travel we are doing, or the size of houses we build, or setting air conditioners to a slightly higher temperature. That would start a revolution. Travel is one of the most significant contributors to carbon pollution, especially if you include business travel, which is unnecessary now that electronic communication can take its place. Also, transportation of products like importing pigs from China, or exporting hay to Saudi Arabia and China, could be curtailed if we were really serious about decreasing CO_2 pollution. But if we resorted to curtailing travel and transportation, the economy would be crippled overnight. Millions of tourist traps would go out of business. Travel has become the most culturally canonized of all our bad habits, and curtailing it would probably cause an epidemic of clinical depression or riots. So instead of reducing travel, our economic geniuses have come up with the carbon tax approach, which allows the rich to burn as much energy as they wish, only then they would have to pay the government to do it.

Simply by operating a farm with more sustainable and bio-intensive methods, substituting human muscle for piston engine power, as Perrine and Charles Hervé-Gruyer detail in their new book, *Miraculous Abundance*, would go a long way toward reducing the problem. With millions more people involved in farming, they would not have time to be hand-wringing about the weather so much.

Another "new" threat turning some paranoids spastic is the Invasion of the Giant Hogweed. (Great title for a movie, and indeed it says on Google that there was a song popular in England in 1971 called "The Return of the Giant Hogweed" by Genesis.) Even its scientific name, *Heracleum mantegazzianum*, is enough to fill a paranoid soul with dread. This invasive weed, creeping toward us out of Europe by way of Canada, originated thousands of years ago in Asia. It is the ancientness of the weed that Americans don't grasp as some of them tremble at videos telling about this grave danger. If hogweed were going to overwhelm us, it would have done so in Europe way before now. The juice or sap in it can blister and scar the human skin and, if it gets into your eyes, might blind you. Scary indeed, but if one just stops and thinks a little, Europe and Asia, where the weed has existed for centuries, still seem to be quite free of any epidemic of blindness. If you rubbed poison ivy juice in your eyes, I imagine you'd be in deep trouble, too.

The blossoms of giant hogweed (I still can't find out how it got that name, but it might be because hogs will eat it, just as sheep will eat poison ivy) resemble Queen Anne's lace but bigger. The leaves are gigantic. Sort of a pretty plant in fact, if I can go by the illustrations I've seen. Herbal history says it was deliberately introduced into France because it makes an attractive garden plant and also is a good honey source. Our Department of Agriculture says that livestock and pigs can eat the weed without harm, so if herbicides won't kill it, fight fire with fire and counterattack with wild hogs. One of the commentators on my website says that the weed came to her country, Latvia, because the Russians brought it there during World War II to grow for livestock feed. I wonder. Another great movie title: "Giant Hogweed Meets Giant Hog."

Our place here in Ohio has been invaded by poison hemlock (*Conium maculatum*), and at first I was alarmed. After all, look what it did to Socrates. I lost two ewes before I knew I had been attacked, but in all truth, I am still not sure they died from eating wild hemlock. I never have been able to control the invasion completely because floodwaters along the creek bring in new seed every spring. The books say it would take something like a half pound of leafy growth to kill a sheep and that it's the seeds that are most toxic. In fact I noticed that sheep will actually nibble a little of the plants in the spring without apparent harm. So I nibbled, too. Even just a tiny speck was so bitter I involuntarily spat it out instantly. I can't imagine anything with a tongue eating more than a little bit of it. As for the fear of children getting poisoned, I wonder how that happens. You can't get them to eat tasty broccoli.

Even if you have managed to escape giant hogweed and poison hemlock, you can't breathe easy. There are so many invasives to keep track of: spotted knapweed, tree of heaven (tree of hell), marestail, water hemp, autumn olive, oriental bittersweet, multiflora rose, and bush honeysuckle to name a few. But so far the world seems to have survived them all. I have long ago quit wringing my hands in useless worry. If you read enough herbals, many plants out there are somewhat toxic to some animals and people at some time in their development and under certain circumstances.

The term *invasive* often turns out to mean "undesirable in my situation." Some people, intent upon reestablishing a natural oak opening landscape in the upper Midwest, consider even maple trees invasive. Maples can grow in the shadow of other trees, and so can shade out oak and hickory trees sometimes. But even oaks, and in fact all kinds of hardwood trees, can be invasive super-weeds in our gardens, which are located next to woodland. They sprout all over like thistles and grow almost as fast as corn. I cut off a black walnut seedling a dozen or so times in the asparagus patch, and it grew back every time. It tempted me to say nice things about Roundup.

Sometimes I like to say nice things about invasives, too. White clover and bluegrass are both not native to America. Neither is the honey bee. A tiny bug that is native to our country is spreading from the West into the entire nation. It spreads a virus called rose rosette disease, which attacks roses. And the rose it likes to attack the most is the invasive wild multiflora. Hooray.

What we are dealing with in modern society is an invasion of too much information that is only partially true or not well understood. It is causing an epidemic of paranoia. Behind the worry is a creepy realization, seldom mentioned out loud, that the most invasive invasive of all is the human race.

CHAPTER 13

One Cow's Forage
Is Another Cow's Poison

"What is food for one is to others bitter poison" is an ancient saying first written down in about 50 BCE. In the Middle Ages it became "one man's meat is another man's poison." The caution apparently applies to farm animals, too. My neighbor called me recently with a strange story. Three of his sheep had suddenly keeled over and within twelve hours were dead. His search for a cause finally centered on a big red maple limb that had crashed to the ground during a storm in the woods to which the sheep had access. The flock had eaten every wilted leaf they could reach. To complicate the matter, my neighbor had thrown some tree trimmings from his yard on a winter burn pile in the pasture. Research seems to indicate that many fresh green tree leaves, when suddenly wilted, contain enough hydrocyanic acid, or prussic acid, to be at least a little toxic. We all know that freshly wilted wild cherry leaves are among the worst. Some sources say red maple foliage can be dangerous, wilted or not, but after years of running sheep and cows under red maples, I never experienced any problems that I could trace to that cause. To add to the mystery, all of my neighbor's flock had access to the wilting maple leaves, but only three were seriously affected. Even more interesting, all three of the dead sheep were related closely and that particular bloodline in the flock had been prone to other kinds of problems.

Pasture plant poisoning is a complicated subject, full of contradiction and supposition, often heightened by the difficulty of identifying plants in the field because of the variation in their colors and leaf forms, the lack of clarity or definition in pictorial renditions and the many colloquial names applied to them. Even the scientific names can change over time. Often sickening occurs not so much because the plants are overly poisonous but because the livestock have access to them when they are starving hungry. Even acorns full of tannic acid have sometimes been a problem in this situation. But my neighbor's sheep were not at all starved but had at their disposal very lush, green pastures because of plentiful rain. Perhaps that was the problem. The sheep had been eating very well indeed, and so perversely (mimicking perverse humans) enjoyed pigging out on something simply because it had a different taste.

I took the mystery to another neighbor who had experienced a freak episode the previous spring. He had found a steer stone dead and bloated one morning in his partially wooded pasture. He could find no really satisfactory cause of death, and finally decided that a touch of frost must have settled on some of the new clover in lower parts of the field just starting to grow. Perhaps the steer had mouthed down a bunch of that slightly frosted forage. But he was never sure. And if so, why were the other steers not affected? When I told him about the three dead sheep and what we thought had happened, he shrugged and shook his head. We had no proof that the poisoning came from the maple leaves, he pointed out, which was true. "There are just way too many possibilities to know the right one without an autopsy, and I doubt even that would be conclusive. There were all kinds of early weeds growing along the creek where my steer died. Who knows if one of them might have been poisonous. But I've been running steers in this field for years without problems."

The neighbor with the dead sheep responded to the poisoning in what might seem to be sort of a hardhearted way, and since he is a very soft-hearted guy, I was a little surprised. "Good riddance," he said. Why? He had experienced a history of trouble with that particular family in his flock, and now he says would not again be tempted to save its offspring even though they showed other good qualities. "Maybe this is nature's way of pointing out weaknesses in a bloodline," he said. Upon reflection, I thought he might be on to something. Nature's so-called law about the survival of the fittest might be as good a way as any to improve a herd or a flock. In one of my favorite books about ranching and grazing, *The Last Ranch*, author Sam Bingham describes in detail a rancher who used this philosophy to select breeding stock. Let nature do the culling, he believed. He bred his cows so they would come fresh in warmer spring weather, even though that could mean a lower price when he sold. He could then turn his herd out to calve on the range, take advantage of the grass from the beginning of spring, and let nature take its course. The calves that survived on their own and came home healthy in the fall roundup would be the ones from which to select replacement heifers.

After growing up with sheep and then forty years of raising them, I had arrived at a similar conclusion, although I didn't always have the heart to follow it. I had reached the point where I wondered exceedingly whether, for instance, trying to breed for triplets was worth it. It just meant often raising one or more of the offspring on a bottle, and I'm not sure that pays. But the instinct of a good shepherd or cowman is to try anything to save every lamb or calf, and there is little doubt that even if triplets mean triple the work, three lambs however small weigh at least a little more at market time than a single. But if I had a big flock, I know that I would have followed the "last ranch" philosophy.

In chapter 12 ("The Invasion of the Paranoids"), I described my experience with poison hemlock. I never had any more problems with the stuff after that one time, even though I could never get rid of it completely. In my rougher pasture along the creek that floods every spring, new plants came in every year, as I said earlier, and

I had to settle for making sure that none went to seed. And yes, I sprayed weed killer after I got too old to do that much hoeing. The books say the seeds are the most potently poisonous part of the plant, so perhaps that is why I had no more problems. I do know the sheep continued to nibble on the leaves in earliest spring because the plants are about the first to grow after cold weather, but I had better things to worry about. Dogs were my sheep killers, not wild plants.

In the final analysis, the ancient saying "one man's meat is another man's poison," or one cow's forage is another cow's poison, is about the best way to describe the situation. If you study the herbal literature very long, you can come to the conclusion that almost everything growing out there has a potential of being poisonous, at least mildly so, under the right (or should I say wrong) conditions. It is not a subject that helicopter parents ought to read about or they might pen their poor children up in the house away from nature (and then have to worry about the paint on their toys). One herbal says that certain people can get a rash like most of us get from poison ivy from handling fresh celery leaves. (And, then again, some people are almost immune to poison ivy.) Potatoes with green skins can be poisonous. Tomato leaves are poisonous, say most herbals, but the deer don't know it so they nibble on ours regularly. I often think that the best way to graze livestock would have them following the example of whitetail deer, now proliferating in farm country. Deer like to eat a little bit of almost everything. In the garden, they taste everything—love tender green growth of raspberry vines, strawberry leaves, sweet potato vines, rose bushes, pea and bean leaves, and corn, of course. No matter how many thousands of acres of soybeans and corn they have to graze on, they come through the garden every morning sampling our sweet corn and snap beans if we don't have our tightwad fence up yet. In the wild, they like to nibble a little of almost all the weeds, even multiflora rose and poison ivy, as well as grass and clover. Somehow they survive just fine over winter, browsing on acorns, nuts, tree twigs, and corn and corn fodder left by the harvesters. In our area, they are increasing in size and vigor and giving birth to twins regularly, although the latest statistics suggest that coyotes are reducing deer numbers. I often have a contrary thought. Maybe we should be just letting the deer proliferate and eat venison instead of beef.

Some people eat hazelnuts with enjoyment; for a few, though, hazelnuts are deadly. Shoots of pokeberry in early spring are almost as good as asparagus, cooked the same way. Later in the year the purplish stems are supposed to be poisonous as are the seeds inside the berries. The berry flesh is loved by birds and chickens, and the seeds pass through them without harm. Horse nettle, in the same family as tomatoes and potatoes, is labeled poisonous in some herbals, but my experience is that horses will eat the weed, thorns and all. I bet that's why it's called horse nettle. And sheep will eat the horse nettle blossoms, which means those plants don't go to seed. Buttercups are poisonous to grazing animals, although I do not know that from experience.

In the absence of absolutely sure and sound information, which is so hard to come by with wild plants, I think we should face the problem of pasture poisoning not fearfully but matter-of-factly, like we do with toxic plant situations we are familiar with. We know that the hydrocyanic (prussic) acid in wilted wild cherry leaves is the same poison that is in rhubarb, in sorghum-sudangrass, and a host of other plants. We don't wilt away in fear from these plants. We know how to handle them in a way that usually avoids problems. Just because sorghum-sudangrass can contain dangerous amounts of hydrocyanic acid after frost doesn't mean we should stop growing it. To avoid nitrate poisoning, you should try not to feed cattle heavily fertilized green corn fodder that has been stunted by drought. For the same reason, make sure your horses are full, not hungry, when you turn them out on lush green pasture for the first time in spring.

The worst "poisoning" problem on the farm is not really from a poison but from bloat—cows and sheep eating too much lush alfalfa or clover too fast or after frost has chilled lush plants. And there are almost as many opinions about how to protect against bloat as there are farmers. And as many antidotes. Today there are antifoaming agents, as they are called, that can be effective in avoiding bloat, which you can find out all about on your search engine or from a veterinarian. They have to be administered every day during the spring and early summer flush of lush legume pastures or after frost in the fall, which can be unhandy with cattle on continuous rotational grazing. (Dairy cattle can be treated at milking time.) Bloating is rare enough that most of us try to avoid it with old-time methods like heavy feeding of dry roughage along with pasture or before turning the animals out on pasture. This doesn't always work, as all the preventative literature will tell you. The time-honored bloat avoider is a pasture that is about half grass and half legume, but that is not foolproof either. What I always did with sheep was to walk among them when I first turned them into a new plot in the rotation, making sure the animals kept walking around while they first mouthed down the lush clover plants so they didn't gorge themselves in those first few minutes. All I can say in defense of that practice is that I never had a case of bloat.

Once, many long years ago, we did have a cow bloat from eating a pile of shelled corn still high in moisture. This was in the really old days when we "shredded" corn, that is, ran the cut bundles through a machine that was sort of like a grain thresher, with rollers sort of like those on an old wringer washer, only bigger. As the stalks of corn ran through the rollers, the ears were snapped off and stripped of their husks on their way to a wagon. The stalks and leaves were shredded and blown out onto fodder stacks. A few kernels were knocked off in the process, and in a day's time amounted to a nice little pile right beside the shredder, to be cleaned up and fed later. The cows were fenced out of the lot where we were shredding, but one contrary old bossy found a way in and she tried to eat that whole pile of corn kernels overnight. I thought she was dying, her stomach area was distended enormously and her breathing was difficult. We got her on her feet, and I jabbed a hole in her side with my pocketknife, right in front of the hip bone where there is normally a kind of sunken area. That's what tradition said to do. The foam hissed out, one of the worst stenches I have ever experienced. We didn't know enough to insert a tube or trocar to keep the gaseous liquids draining out, but somehow the cow actually lived.

The problem is that possible poisonous wild plants are bound to increase as pasture farming becomes more widely practiced. This will be especially true of land now being allowed to grow back to brush, because it is not profitable for large-scale grain farming, but will shift back to pasture when that becomes profitable. No doubt plants like the very poisonous white snakeroot, which likes a woodsy habitat, will become problematic, as they were during the first land clearings. A neighbor of my childhood, standing over a cow he said died of eating white snakeroot, told me that even the cow's milk was poisonous. I can't say whether that's true from experience, but a child never forgets being told something like that.

Quit worrying so much, I say. Animals by and large are smarter than we are about wild plants. In almost all cases, poisoning comes not from ingesting a little but from too much, and animals normally know enough not to do that. And in some cases a little bit of a toxic plant can actually be beneficial. If you must worry, concentrate on loose-running dogs, which unfortunately don't eat poisonous plants.

~~~~

CHAPTER 14

Pasture Farming as Part of Garden Farming

O nly a relatively few new age farmers have a cultural connection to pasture farming compared to the number who are familiar with gardening. Even for those of us who grew up and have lived on farms most of our lives, modern methods of raising animals almost exclusively on pasture forages is a new idea—or rather a very old idea. Somewhat disbelievingly myself, I spent thirty years experimenting with rotational grazing with eight very small plots of an acre each. I convinced myself that even at this scale, or perhaps especially at this scale, pasture farming works. The know-how involved has advanced considerably since I wrote a book about it, *All Flesh Is Grass*, and most of it assumes the attitude of industrial farming—that is, see how many animals you can cram on a given pasture to make the most money. Nothing wrong with that, but rotational grazing is even better suited for the part-timer or homesteader who wants to make a little money, maybe, but who is mostly interested in an easy way to produce healthful meat, milk, and eggs with the least amount of time and money. For them, if pasture farming produces tasty, healthful food and pays the taxes and upkeep on the meadows, that's good enough. The added benefits, that is, the beauty and wildlife diversity that pastures bring to a homestead, plus a place for a pond or a hill for sledding or honey-grazing for bees, or room for solar panels—hey, that's profit too. When you think about all these free gifts from pasture farming that you don't have to declare as taxable income, casual rotational grazing begins to look as profitable as a no-nonsense commercial project. But much of what has been written focuses on the commercial aspects of rotational grazing rather than the fine points that can make it a low-cost, low-labor endeavor. So here's a different take on it.

For the uninitiated, rotational grazing is a kind of livestock and chicken farming that mimics nature much more closely than confined animal facilities. Overhead is low, CO_2 pollution low, energy expenditures low, erosion almost nil, and no annual cultivation is necessary. About the only machinery you need is a mower, to cut down uneaten weeds after the grazing animals are moved to another plot in the rotation. You might well be able to do the mowing with a large lawn mower. Rotating the animals from one pasture to another means they always have fresh grass and clover. The casual grazier should understock his pasture—that is, keep fewer animals than the land could actually support so that your animal(s) rarely need any other food except in winter. Where winters are not severe, animals can find some grazing almost all year round.

When to move your animals from one plot to another is almost the whole art and science of this kind of farming. But the main reason for doing it, at least for the non-commercial grazier, is not to see how you can make use of every little blade of grass or leaf of legume profitably, but to keep the animals from going back for at least a month and a half to the plot they have just grazed down, in order to reduce the population of internal parasites on that plot. It is not necessary for the casual grazier to follow the dictates of science and commercial farming to gain the most possible pounds of meat, milk, or eggs from your pasturing. You don't have to measure the height of the forage with a ruler or base decisions on one of those gismos that tells you the sugar content of the forage, or the results from another gismo that determines the protein content of the clover. The best advice is to treat your pasture just like you do your lawn to maintain a good stand of grass. You need to know the pH of your soil. Clover, an essential part of rotational grazing in my opinion, to do well should have a pH of 6 to 7. But breaking the life cycles of internal parasites, which hatch in the soil or manure and crawl up the plants and are consumed by your animals, is absolutely critical. Try to keep your animals from coming back to the plot they are presently grazing for around forty-five days. That keeps the parasite population low. Messing around with worming animals is a pain, especially these days when internal parasites are becoming immune to wormers.

Some refinements of the grazing art for casual graziers are not emphasized as much as I think they should be. There are all kinds of forages being brought into rotational grazing, all with various pros and cons for the most progressive commercial grass farmers. But for those not constricted by the need to make a good money profit from it, three old-timers should be a big part of your forage program: red clover (or crimson clover in the Deep South), white clover, and bluegrass. White clover and bluegrass don't get much good press because they volunteer and grow mostly on their own, meaning that seed dealers can't make much money from selling them. But they are easy to maintain forever and can stretch the grazing season later in the fall and earlier in the spring with less cost and labor than other forages, especially where the goal is subsistence feeding, not trying to produce the highest possible commercial gains. In a milder winter, like what we experienced here in Ohio in 2015–16, the bluegrass and white clover survived right into winter, and our chickens were grazing both species into January and bluegrass right on through February until new growth in spring. Both species were especially vigorous around barnyard buildings where rainwater off the roofs kept them watered well enough to supply a few hens with good grazing even in droughty August. For those who keep backyard chickens, this is important information: A well cared for bluegrass / white clover lawn can supply most of the hens' ration through most of the year, even in winter when snow does not cover the ground. Also, when you mow your lawn, let the clover dry like you would when making hay. Then rake up the dry clover and store it out of the weather to feed hens in winter. If you splurge and use sprinkle irrigation on small plots as you would on your lawn, you can graze bluegrass and white clover right through late summer dry weather. Or save the second growth of a red clover plot for late-summer grazing. It is not as drought-resistant as alfalfa, but almost. If you are thinking about making hay from red clover, a new variety, Freedom, has hairless stems, so it dries quicker, says the Byron Seeds catalog. This company also makes a point that's seldom heard: Red clover has a bypass protein level twice that of alfalfa, and new varieties are higher in digestible neutral detergent fiber (whatever that is) compared to alfalfa. Therefore, "not only can a cow derive more energy from red clover, the protein is more valuable, too," according to Byron Seeds.

But my reason for championing red clover is that it will sprout from broadcast seeding better than other clovers. That means you can plant it with a low-cost, hand-cranked seeder looped over your shoulder almost any time of year. However, it is better to do it early in spring when the soil is still a bit frozen on the surface. It's best to sow on mostly bare ground, but you can plant right on top of existing old forage growth and get a fairly good stand if you increase the seeding rate to about fifteen pounds per acre. You may not get really professional stands this way, but they're good enough for casual grazing, especially where white clover and bluegrass are already established. A rancher in Kansas told me years ago how he would save a field with a good stand of red clover for winter grazing. His cows would graze the dying clover almost as well as living green foliage in late fall, and trample seedheads into the ground, which, along with the clover seeds that passed through the animals in the manure, reseeded the field quite well.

The biggest expense and the most necessary one in rotational grazing is good fencing. You can avoid spending a lot of money by using electric fencing, and many rotational graziers do, but if you listen to any of what I'm saying, do not rely on electric fencing around the outer perimeter of your pastureland, no matter what the experts tell you. Livestock invariably get through it, especially, it seems, when you are away from the farm. If your cow causes an accident on the highway, think what that might cost you. Use high-quality woven wire fencing around the perimeter, damn the cost. This fence also discourages roving dogs, and if you are raising sheep, dogs are your biggest problem. Better yet, if you want to be done with fencing for the rest of your life, and especially if you have horses that inevitably ride down woven wire, use wire panel fencing, damn the cost twice. It is more expensive than woven wire, prohibitive for a large pasture farm if you don't figure in the fact that it will last your lifetime. It comes in sixteen-foot lengths and varying heights. Get the tallest, which are about five feet high. The extra advantage is that you can take a section down anywhere to gain entrance to, or egress from, the field or to mow and clean up brush growing up in the fence.

Electric fencing comes in handy, though, inside the perimeter fence to separate the various paddocks and allow you to change their size and number as you figure out what is the best rotation and plot size for your circumstances. Electric fence is almost essential for commercial strip grazing where the fence is moved every day, or even more frequently, to give the animals fresh grass continuously. The casual grazier is not usually going to want to do this, so he or she should lay out the paddock sizes more or less permanently, and the animals are moved as necessary to fresh grass. Once I figured out a schedule that suited my purposes, I put in permanent woven wire fencing around the eight paddocks, mostly to discourage roving dogs even more.

The negatives in pasture farming are getting water to every paddock, keeping the grazing animals from trampling and tearing up the sod in wet weather, and persistent weeds that the animals won't eat. But smaller-scale graziers can handle these problems fairly well. Since they have only a very few animals, trampling is not as critical a problem. Since they are not intent on making every last possible penny, they can keep the animals penned inside and feed hay during the critical mud days of March. They can also arrange paddocks so that providing water is not laborious or expensive. Sometimes they form paddocks in pie-wedge shapes so that they all narrow down to the barn at the center. That way, rainwater off the barn roof will supply water to any plot most of the time.

As for persistent weeds, on a small acreage they can be hoed out by hand. It is not as impossible a job as it might seem on an acreage of two to twenty acres. Two of my sisters, who are not young anymore (see chapter 21, "The Homebodies") have done it on their pasture for half a lifetime now. Or use an organic or chemical weed killer, depending on your philosophy. Most weeds will slowly go away with persistent mowing and grazing. Canada thistle was the worst weed I had to contend with, and I learned that mowing religiously after every grazing slowly diminished them. When a plot containing Canada thistle was grazed a second time in the same summer, the sheep would eat off the tops of the new growth of thistles as the flowers budded. Then when I mowed off the rest of the half-dead plants after moving the sheep to another plot, the thistles grew back only slowly, and after a few years not at all.

A weed that exasperated me very much is what we call barnyard millet. It is a grass that comes up in dry late summer. Livestock will eat it—grudgingly—until it goes to seed, which it does quickly. After it seeds, it looks like it is blotting out the good forage. I finally gave up on it in despair and thought about plowing up the plot where it grew particularly well and planting a new sod. But with persistent, timely grazing and mowing, the millet just started waning of its own accord. Summers of plentiful rain, which millet does not like, seemed to curtail its growth. The other grasses and clovers won out when there was plenty of moisture.

Pasture farming the garden farm way has other possibilities as well. Chickens and even sheep and goats can be introduced into gardens or hoophouse plantings to help with insect and weed control. Geese have long been used to weed strawberry beds. *Miraculous Abundance,* the book cited earlier, has photos showing how the authors incorporate chickens right into their greenhouses.

In any event, walking your pasture in spring when the lambs and calves bounce over the turf and meadowlark song drifts over the grass toward you is about as close to heaven as you can get.

CHAPTER 15

~✦~

The Wild-Plant Explorers

For those of us who are following the motto of "get small and stay in," there is a way to acquire all kinds of delicious food or attractive landscape plants, especially tree fruits and nuts, without spending any money at all: from wild plants. Gathering them makes an adventure out of every walk through the woods or drive along a lonely country road. And propagating them on your homestead, though a gamble, can often have gratifying results or even get you started in growing and selling an unusual, profitable product. For instance, pawpaws have become a commercial crop in Ohio, and no doubt elsewhere, in recent years.

We have a pear tree in our front yard, which is also our orchard, and we're not exactly sure where it came from. I know where it came from *sort of*, but when you are a wild-plant explorer, knowing precisely where you discovered something does not necessarily mean much. Years ago, I found a pear tree growing wild in the woods. How it got there I don't know. I think it was an "escape," as wild-plant hunters say, from the property next to us, but it could also have come from seed carried by wildlife from the orchard that once grew just across the road. Pioneer settlers, as part of the deal to get their free homesteads in Ohio from the government, were obliged to plant an orchard. On the oldest maps, the locations of these orchards are marked, and for a wild-fruit-tree hunter they can sometimes explain the origin of a mystery fruit tree found growing in old woodland or along country roads or in fence rows. At any rate, this particular pear tree was dying from too much shade of taller trees, so I took seeds from it and, as I often have done, stuck them rather haphazardly (I am sort of a wild planter, too) in the ground in the corners of a cold frame, just to see if they would sprout. When a seedling did come up, I thought it was a cherry tree, since I had also planted cherry pits there, from an heirloom tree growing in a nearby village. I transplanted the little seedling to the orchard. The leaves soon enough proved not to be cherry, unless it was of some new strain unknown to mankind up to now. I decided it was an apple seedling since I was always sticking apple seeds in the ground, too. But in a few more years, when the tree was about twelve feet tall, it bore fruit. Pears!

The odd thing about the tree is that the buds on the ends of the twigs are as sharp as thorns. Perhaps I have some weird kind of throwback to prehistoric times. The books say that pears are a very ancient fruit, originally growing wild in the forests. Some three thousand varieties are recognized around the world. My sister joked that ours is a prickly pear. The fruits are average in size, more roundish than pear-shaped, dull yellow when ripe with a faint blush of red on some of them. They are a little sweeter than those of our Bartlett and Seckel trees growing nearby. This was most surprising because the fruit of the wild pear in the woods had not been nearly as tasty. We have never sprayed the tree, which is probably why occasionally some of the fruits are small and knobby. I began to fantasize about establishing wild pear forests all over the country like the apple forests that still exist in Kazakhstan. Maybe a tree would come along with juicy, sweet pears on it as big as watermelons . . . well, you know, little watermelons.

I have to thank the North American Fruit Explorers (NAFEX) for showing me how to make life on the land even more exciting than it normally is by hunting for wild foods good enough for the garden farm. This organization started with a handful of members in 1967, and I became acquainted with them a few years after that.

They were a charming and rather contrary bunch from all over the United States. They taught me mostly how the quality of wild fruits can vary dramatically. Their correspondence and their journal soon taught me tidbits of fruit lore I had not known. For instance pawpaw and persimmon trees grow naturally this far north. As a result, our property, in addition to our strange pear tree, has become home to pawpaws, wild apples, wild peaches, hickories, a butternut, a shadbush, a dogwood, red cedars, a catalpa, a sycamore, a hazelnut, and a chinquapin oak, whose acorns are mild and quite edible. Most of these we brought in from surrounding natural areas and a few from other states. I have had a few failures—a coffee tree seed from Kentucky didn't sprout, and the persimmon hasn't fruited because I didn't know I needed two trees for pollination. Maury Telleen, the founder of the *The Draft Horse Journal*, once sent me several hickory nuts from his Iowa farm that were as big as eggs—bigger than any of the regular "bullnut" hickory species that grow here. I planted the nuts in a fence corner where I could easily protect them from sheep and cattle by blocking off the area with a gate. One of them sprouted and I was very excited. But it died. Another of the original nuts that I gave to a friend is flourishing—about eight feet tall now. I planted mine in old farmed soil. She planted hers in uncultivated soil, which provides even more evidence of my observation that trees grow better on virgin soils than on long-cultivated land.

There are about a dozen old roadside hickories in our neighborhood that bear choice nuts. We think the earliest farmers saved these trees when the land was cleared because they knew a good hickory when they found one, with thin-shelled nuts easy to crack into "whole halves." We call our favorite one Francis because it grows near to the home of a farmer by that name, now passed away. Last year the nuts from Francis literally covered the ground under the tree, and we picked up over a half-peck of them in less than twenty minutes. I have several trees started from Francis in our orchard. But because wild hickories do not always come true to the mother tree, none of them have nuts as nice as the old matriarch. Ours have thicker shells and are harder to crack. But the nutmeats are larger than normal, so I consider it worthwhile to have planted them. One thing about hickories: success comes more often from starting them from nuts than to try to transplant seedlings. Many tree seeds need to be "stratified" before they will sprout—that is, submitted to a period of freezing weather. You can do that in the deep freeze or you can mimic nature and plant very shallow in the fall and let winter weather do it. Where squirrels or other wild animals are a problem, laying a piece of metal screening over the planted nut over winter can help protect it.

Our countryside is dotted with wild apple trees, along the back roads, in fencerows, and in cutover woodland. Sometimes they are little better than crabapples, and sometimes they are as nice as named varieties. Folklore says Johnny Appleseed planted them. He must have been a very busy planter indeed. One day when I was roaming along an overgrown old fencerow (great places for a wild-plant hunter to roam), I came upon two old trees laden with marvelously red apples. Where a farm disk had cut into a root in a previous year, little sprouts had grown up and put out their own new roots. Here was a special opportunity. If I could transplant such sprouts, it would mean preserving the parent tree just as would be the case with a graft. I went back home for a shovel and an axe, cut off two of the sprouts from the main root, and carefully dug them out of the ground with their own little roots intact. Both grew fine next to our shop. They are not as good an eating apple as our Winesap and Grimes Golden, but they make excellent red applesauce and pies and bear heavily (too much so) every other year. We have never sprayed them. Sometimes some of the fruits are wormy or touched by apple scab, but there are always plenty that make good pie and sauce with just a little paring off of rotten spots.

If you are buying apples from the grocery store, rightly enough you expect that they aren't blemished or wormy. But with free fruit, a bit of imperfection is not so objectionable. While peeling an apple, cutting away bad spots is only a little more work. So in a way it is because of money that a whole bunch more money needs to be spent unnecessarily, spraying fruit to keep it free of blemishes. The same holds true with the trees themselves. By buying trees of named varieties, you are assured of quality fruit, and in the long run, or rather the short run, that may be a smarter way to get it than hunting wild trees with only a slight chance of finding specimens as good as those from commercial nurseries. But purchased fruit tree varieties (and ornamental trees) are becoming quite expensive these days, which leads wild-fruit explorers to decide to take their chances on wild trees or planting seeds instead of buying nursery seedlings because, not having to spend money, they are more willing to put up with lesser quality. Besides, it's more fun.

Another wild-apple adventure happened when we were visiting relatives who had just bought a little farm about ten miles away. They had invited us over to "show you something." A bit of a ravine ran through their place, which had at one time been grassy sheep pasture but now was overgrown with trees and brush. As we walked our way through it, we began to encounter apple trees laden with fruit, and in some cases quite attractive fruit. I could hardly believe my eyes. Our relatives were laughing because they had only recently discovered the trees themselves and they knew I would be excited. When we reached the end of their property and the wild apple grove, we were not far from an old homestead and orchard. We reasoned that in bygone years, the owners of that orchard fed surplus apples to their livestock, or perhaps made cider and fed the pomace to the animals, who in turn spread the seeds in their manure in the pasture. But who knows. I like the idea of Johnny Appleseed doing it, even if he would have had to be well over a hundred years old to have been around when these trees sprouted. Mystery makes life more interesting.

Peaches are still our most successful venture in growing seedling fruit trees. I became friends once with an old and masterful gardener in a nearby village, drawn to him by his rose garden, which he kept almost as meticulously neat and beautiful as the one behind the White House. He had a peach tree he called Lemon Free. The flesh was a lemony colored freestone, hence the name. He gave me a couple of fruits from it one day and said, almost mysteriously, "Plant the seeds. Lemon Free comes true to the mother tree." To this day, I don't know if that is completely true or not, but I was of course fascinated. I planted the seeds on the edge of our tree grove behind the house and promptly forgot about them, which is typical of me. I should keep a daybook. About five years later, I noticed a little tree with peaches on it growing at the edge of the grove. At first I wondered where in the world it had come from. It took me about two days to remember, helped along by Carol, who never forgets anything, especially things I tend to forget like wedding anniversaries. And yes, it was just like the Lemon Free tree of my friend. To this day, we are eating these peaches because, although we have tons of better ones from other trees, these come late in the season, in mid-September.

Before we had our own peaches, we purchased Red Havens for eating and canning and threw the skins and pits to the hens to clean up. Soon, little peach trees were coming up all around the chicken coop. Curious, I let them grow. In about four years they began bearing fruit. About one of every three trees bore peaches of good flavor and only slightly smaller in size than store-bought peaches. I cut the ones down that bore only poor fruits. We have rarely been without peaches since then. More seedlings continue to come up from pits wherever there's enough sunlight, and I have learned that the trees are sort of self-renewing. If a large branch breaks off, as has happened when it is laden with fruit and I didn't get around to putting a prop under it, the tree sends out new branches that bear in a year or two. On one tree, the main trunk is starting to die but a new shoot is coming up from the roots to take its place. I had never read about this advantage of peach trees, but checking the "literature" I learned that there is a strong opinion among many horticulturists that trees grown on their own roots rather than on grafted rootstocks grow more vigorously and live longer.

Finding the chinquapin oaks from which I got an acorn to start my own tree was an adventure in itself. When I first came across the trees, on land owned by our daughter-in-law's family, I could not identify them at first. They were huge, ancient trees, with leaves similar to chestnut oak but longer. The name *chinquapin oak* is deceiving and even books on trees tend to confuse it with the real chinquapins, which are related to the chestnuts, not the oaks. I tasted the acorns. They were quite mild, much better tasting than any other acorns that grow here. They are in fact not native here, exactly. Tree historians think that Native Americans spread chinquapin oaks wherever they moved because they are quite edible, like the California white oak acorn, which was a principal food of West Coast Native Americans for centuries. So now I have one started in our yard. Deer have nibbled on it twice, but it still grows—about four feet tall now.

I suppose I could say that my open-pollinated corn is a wild plant, too. It goes by the name of Reid's Yellow Dent, and I got it from a farmer in Iowa, but it rarely comes true to the mother plant, so having a name is hardly practical. I've saved seed from it now for forty years and still, when I plant a kernel, I can't predict what will come of it. It might have a normal ear or a nubbin or no ear at all. Sometimes the ears are gigantic—fourteen inches long with twenty rows of kernels and sometimes nine inches long and twenty-eight rows of kernels. The fatter ears of medium length invariably have more grain on them by weight than the longer ears, mostly because there are more rows of kernels on the fat ears and the kernels grow deeper into the cob than on the long ears. For practical, commercial production, I would be better off growing hybrids, maybe, to avoid all this inconsistency, but I prefer the adventurous possibility of finding a giant ear someday, maybe eighteen inches long with thirty rows of kernels. Who can say? When ancient farmers in Central America started growing maize, the ears were scarcely three inches long, and if some dreamer had said that they would stretch out to 12 inches or more some centuries later, the Aztecs and Mayans would have howled with laughter.

Variation within the same species extends even to characteristics like fall leaf color and sugar content in the sap. Maple sugar producers save seed to plant new groves from trees whose sap tests higher than average in sugar content. Some NAFEX members believe that fall leaf coloring can vary even within the same tree species. A stroll or drive through old residential areas of almost any village here in Ohio reveals old maples whose extravagant colors seem to have little to do with the variety involved. I like to think these trees were deliberately propagated from seed or seedlings long ago.

Sometimes planting tree seeds from the wild is the easiest way to get odd varieties started on your place. I think sycamores are among nature's most strikingly beautiful trees, but they are difficult to propagate. I put seed balls picked directly from a tree along the river a few miles away into pure sand in a glass tank originally used for a fish aquarium. I kept the sand moist and put a piece of clear plastic over the top of the aquarium to maintain a moist environment inside. The seeds all came up, and one of them is now a tree outside my office window nearly fifty feet tall.

As you collect seed from your plant treasures, do keep a record of where you plant them or you will end up like me being surprised occasionally—more than you should be. Walking through one of our tree groves last year, I found some butternuts on the ground on one of the driveways we maintain through the trees. I was totally taken aback. I've known that patch of woods for nearly eighty years and had never found a butternut tree growing in it. How could this be? It took me three days of wonderment to remember. In one of my wilder moments, I had planted butternuts right there along that pathway right after we had purchased the woods.

Chapter 16

The Most Stubborn
Farmer of Us All

W hen I say that, to succeed, farmers need to be very
stubborn, I think of Uncle Cranky, as he was known
locally. He had a reputation for a contrariness so
unyielding that newcomers to our county sometimes mistook it
for saintly patience. Only once did he meet his match, and then
in a form neither human nor animal, which is a remarkable feat
since there were plenty of examples of both in our neighborhood
that would win the grand championship of obstinacy anywhere
they would not have to compete with Uncle Cranky. As difficult
as it might be to believe, his match in mulishness turned out to be
Kieffer pears.

As the story has come down to us in local folklore, Uncle
Cranky had little use for his neighbor, Old Brown, and had a habit
of muttering something that sounded like "shiftless hilligan" when
he passed the latter's farmhouse. While Old Brown's haphazard
farming could provide several good reasons to justify such muttering, it was the pear tree in his yard that riled Uncle Cranky the
most. Every year several bushels of pears went to waste under it.
Uncle Cranky could not abide waste. His wife said he liked his
soup extra thin so no food particles stuck to the sides of his bowl.
One day when Old Brown was complaining about the price of

corn, Uncle Cranky could withhold his burning, pent-up criticism no longer. "Whilst you're complainin', you let a whole year's worth of fruit go to ruin on the ground," he growled.

"Them things?" Old Brown laughed. "They ain't fruit, Cranky, them's Kieffer pears. Hardly fit for hog feed."

Cranky snorted right back. "You think your granddaddy planted that tree for the fun of it? You just never learned how to handle 'em."

And so, as the saying went in our community back in those days, "the 'gantlet' got throwed." Grabbing a corn fork, Old Brown scooped up a bushel or so of Kieffers and set the basketful in the bed of Uncle Cranky's pickup.

"There now," he said, savoring the moment. "Let me know when you have learned how to handle 'em well enough so's a body can eat one of the stony things."

Uncle Cranky did not understand winter pears, let alone Kieffer winter pears, but he did not yet understand that he did not understand. Back home, he laid the green, rock-hard fruit on shelves in the cellar to age awhile. In a few weeks he cut one open, or tried to. It was like carving on a cue ball. Finally he clomped it into halves with an axe on the splitting block. He could just barely bite into one of the splits, although the fruit looked white and juicy. Just needs a few more weeks to soften up, he told himself.

Each day, he checked his Kieffers. It became a sort of ritual. He would pick up a pear, bite into it, or try to, then put it back on the shelf and stare at it stonily for a few moments. Just as stonily, the pears stared back. November passed. The pears took on a yellowish hue. A few brown specks appeared on some. But they softened on the surface only enough to curl a particle under pressure from a fingernail. By December, he could discern a slight softness to the surface of the pears if he mashed one in the vice. It squished a little. He began to mutter, at the end of his staring ritual, a soliloquy on whether or not Kieffer pears might best be used to throw at loose-running dogs.

In March, he axed open another one. It was flecked with brown, apparently rotten although of a rather unyielding rottenness. That did it. Uncle Cranky snapped. "You stony devils think you have beat me, dontcha," he growled, his voice rising. "Well, we'll see about that. You're going to get fed to the hogs now, think you're so smart. To the hogs, damn you." Out the cellar he marched, bearing his basket of pears, pausing only to make sure Old Brown was not passing on the road, and heaved them over the fence with grim relish. "There now, we'll see who wins this little fight."

The hogs came running and they indeed seemed to want to eat the pears, or at least liked the *idea* of eating them. But nothing in their upbringing had thus far prepared them for such hard and ungainly victuals. They did not yet know how to press a pear into the ground with their snouts and gnaw off little bits. As the immortal words of Uncle Cranky's wife have come down to the present day: "One big ole shoat finally kotched one a' them pears up in its jaws, tried to bite through but couldn't, swallered it anyway or tried to, and choked to death right there in front of us. Oh it was a sight."

Gardeners who have had to deal with Kieffers doubt this old story only a little. The more knowledgeable among us joke that there are ways to soften the fruit—first printed in a magazine, *Hard Farming*, in 1883, we joke.

1. Allow pears to age on the ground until December.
2. Spread fruit on granite slabs or similar hard surface and beat with hickory maul.
3. Boil until soft or for three days, whichever comes first.
4. Can and serve five years later.

Another way is to load up one of those old Civil War cannons with as many Kieffers as the barrel will hold and blast them against the abrasive wall of an old concrete silo. This will both skin and soften the fruits at the same time, or knock the silo over.

Uncle Cranky is dead and gone now but not the Kieffer pear tree that so riled him. A young lady of my acquaintance now pretends to rule over the tree, and she is a new and possibly even more contrary farmer than he was. She shall remain anonymous because she cans Kieffers. Any pretty girl who can soften the heart of a Kieffer pear is a positive threat to the morals of the surrounding community. But she protests. The pear really isn't that hard, she claims. "Makes good preserves, and is the only pear tree I know that produces every year without any care at all."

And so the Kieffer, as a variety, goes on living into a second century at least. Fireblight kills more amenable varieties like Bartlett. Pear psyllid, which on occasion threatens to kill off the whole West Coast pear business, is too smart to attack Mr. Kieffer. Only the most resolute fruit worms will try burrowing into one. Pear slugs riddle the leaves of Bartlett while ignoring Kieffer. So you don't have to spray the trees. They will grow anywhere from the coldest regions of fruit production to the warmest. They live to an old age, and even in death the wood can be used profitably. The most expensive carpenter's hand planes and other tools are made from the heartwood of pear. Woodcarvers like it, too, especially large blocks of it that are only available from big old trees. If all else fails, the wood makes a hot and aromatic hearth fire.

Despite their contrary hardness in the early fall, Kieffers actually do soften enough to be ground and pressed for perry—pear cider—just like pressing apples. Perry has often been mixed with apple cider. The old-timers thought that apple cider with one-tenth portion of perry made the best combination.

Another advantage of winter pear trees, seldom mentioned, is as ornamentals. If you don't mind those fruits on the ground in the fall, the trees bloom marvelously in the spring, and in fall until very late they decorate the landscape with shiny coppery-gold foliage even in the northern parts of the country.

To plant a Kieffer, drop a pear where you want a new tree and run over it with a bulldozer. Or you can cut the seeds out of the fruit and store them in a plastic bag in the refrigerator or freezer to stratify them for spring planting. The trick is to not let the seeds dry out.

If you like pear taste, or even if you don't, try the Kieffer. A Kieffer in hand is worth two Bartletts in the bush, especially when a loose-running dog is crossing your property.

CHAPTER 17

Have We Deflowered Our Virgin Soils?

One of the many gems I have learned from my farmer cousin, David Haferd, sticks in my mind. We were looking at his wheat field one day in about 1985, and he asked me if I could see anything unusual about it. Well, yes. One side was a bit taller and greener than the other side, now that you mention it. Probably planted on different days or with different amounts of fertilizer or different varieties, I guessed. He shook his head. "Nope. That more robust stand is on land that was cleared about forty years ago. The other is on land cleared eighty years ago."

I was at first only bemused to know that there were farmers still alive who knew the history of their farms that well, or who could remember such details. But then the full import of what he was saying hit me. The logic seemed inescapable: The soil cleared and farmed in the 1940s had more natural fertility left in it than what had been farmed since 1900. That would suggest the most contrary question a farmer could ask today. Could it be that all of our great advances in scientific farming were in reality steps backward in soil fertility? The question was especially pertinent because David farmed very carefully and kept to a rotation that always included green manure crops. In most of those years the land received a goodly dose of animal manure, too, as well as some chemical fertilizers. Since he helped clear the forty-year-old field, I asked him if it was any different, in the first years of cultivation, than the eighty-year-old field. "Oh, yes," he said, "it worked beautifully, very loose and loamy for a few years." He could not remember if there were any differences in yield, but since, as time went by, he added more fertilizer, a proper comparison was not possible.

It was not until I settled on my own little backwoods farm that I began to wonder if there could be anything rare and precious in virgin soils that had been lost on heavily farmed soils. About five acres of our farm is old-growth forest from which a few trees have been harvested over the years, but the land never cultivated. We built our house on the edge of it in the shade of 150-year-old oaks and built the barn in a clearing we made within the trees. The land outside the tree grove had been farmed hard for over a hundred years. I have thus been able to observe closely how plants grow on both kinds of land. Believing as I do that there is something almost magical about soil, I must be careful how I express my observations, because I do not have much of a scientific background to warrant what I want to say. I will have to lean heavily on my observations and the subjunctive mood.

One difference between long-time farmed soil and virgin soil seems unquestionable. Trees start in virgin soil much easier than in the farmed soil. I noticed the difference most of all with peach trees. I first planted several varieties on the old farmed soil. They grew only slowly—two not at all—and did not thrive. Only one survived, and it is half dead now. On the other hand, we threw skins and pits of peaches we bought when we had none of our own to the chickens outside their coop in the tree grove. To my surprise, many of the pits sprouted and grew vigorously. The books say that seedling peaches do not come true to type from seed and may not produce quality fruit. But I let my volunteers grow anyway just to see. After all, they were free.

They grew remarkably fast—head high by the end of their second year. In the fourth year they produced fruit, almost as large and tasty as the Red Haven peaches whose seeds I had thrown out on the ground in the first place. In the fifth year and then continuing to the present, they produced quite well, with more trees coming up every year from the pits we discarded. I transplanted seedlings occasionally, but there was no need. The seeds sprouted easily and soon overtook the transplants that I had planted on long-cultivated soil. I have to think this can only be explained because of the virgin soil. I also believe the big hardwood trees surrounding the barn and chicken coop clearing protect the blossoms somewhat on frosty nights, but could something in the soil also be giving protection?

I have, just for fun, thrown apple pomace from cider making into the woods. Many of the seeds sprout readily. They don't last long because there's so much shade. When I scatter pomace out on fields that have been farmed for years, I do not get anything like that virgin soil response.

Another example that seems magical is the way weeds grow in woodland soil if given enough sunlight. In my little cleared barnyard, giant ragweed, burdock, and lambsquarter, among others, grow with almost unbelievable vigor to great size. These weeds, of course, grow mightily enough in farmed-out soil, but in virgin soil I can hardly describe their growth too dramatically. Already by the middle of June, the bottom leaves on the burdock are big enough to wrap a baby in. The giant ragweed and lambsquarter start slower, but by August, they are taller than the chicken coop nearby. I'm talking fifteen feet for the giant ragweed and twelve for the lambsquarter. If they had full sun, I believe they would grow taller. There is no fertilizer of any kind applied to these plants. No fine seedbed is prepared for them. Other weeds grow around them. Their great growth is accomplished without the aid of any of what we call modern cultivation methods. Needless to say both giant ragweed and lambsquarter seed are very high in protein. Grazing animals love the leaves of both. Lambsquarter makes good salads. Many people regard burdock root as a tasty vegetable. Some herbalists consider it to be good medicine for a variety of ailments. My question: are the health benefits possible because burdock comes most of the time from uncultivated woodland soil?

Ginseng offers persuasive evidence that the answer might be yes. Ginseng grows wild over most of the eastern U.S. The price on Asian markets is so high that growers here go to great labor and expense to produce domestic ginseng in slat-covered beds to imitate the shady environment of wild woodland ginseng. But buyers will pay a lot more for the wild kind. Obviously, they have ways to tell the difference or traders would try to cheat. Therefore the ability to identify wild ginseng by its health effects can't be just superstition, can it? Is it possible that ginseng grown on wild virgin soil really does have a more healthful effect on the body than domestic ginseng?

Native hardwood trees sprout and grow in my virgin soil with awesome vigor when they get enough sunlight. The maples don't even need much of that. The tree literature abounds with stories urging people to plant more trees, while on our place we are being invaded with trees to such an extent that we have begun to refer to our grove as the Green Monster. Our biggest gardening task is weeding seedling trees out of our gardens next to the grove. If we did not mow the lawn, the saplings would soon advance across the yard and overwhelm it. We put our raspberry patch next to the woods, since black raspberries can stand a little shade. The ash and walnut seedlings that spring up in the permanent leaf mulch of the berry patch grow five feet in a year. If I cut them off at ground level, they grow back that tall in half a year. It is just awesome. To get rid of them, I have to either dig out the roots by hand or spray them. Wherever sunlight can penetrate the forest floor, thousands of seedling ash, maple, elm, oak, hickory, and walnut appear as if by magic. That's why I argue vehemently against the notion that white ash trees are being destroyed by the ash borer. On my place all the older ash are dead, but a whole new generation has sprung up—just as happened when we thought the elm trees were a goner from Dutch elm disease. The beautiful little redbud tree we introduced turned into an invasive pest. The pawpaw tree I planted looks like it is going to take over the whole eastern side of the grove unless it loses the battle with that other awful invader, white mulberry. Notice that I am bad-mouthing plants that have proven food value for animals and humans. Maybe I should be imitating southern farmers of the past who kept mulberry, pawpaw, and persimmon groves to fatten hogs.

Science explains the vigor of plants in virgin soil to some extent. Plant roots commonly are covered or infiltrated with mycorrhizal fungi, which in symbiotic relationship with the roots allow plants to grow more healthfully, taking advantage of minerals, especially phosphorus, that might be present but otherwise unavailable to the plants. Virgin soils typically contain lots and lots of these growth fungi, if I may call them that. Various mycorrhizal preparations are now on the market as soil enrichment additives, attesting to the fact that we have farmed much of the natural population out of our soils.

I have searched the agronomic and microbiological literature for forthright comparisons between the fertility of virgin soils vs. farmed soils. The effort left me bleary-eyed. The language of soil scientists is steeped in very long words, and the conclusions flowing from these studies use the subjunctive mood as much as I do. In most such studies, virgin soils appear to have larger and more diverse populations of microflora and microfauna, and especially more worms and bugs and visible fungi than farmed soils, but I haven't found any study yet that flatly states that virgin soils are superior for growing food crops than soil farmed for a long time with chemicals. One reason for this hesitancy is that only a few scientists want to go against the grain of modern agronomic orthodoxy, which teaches that most soils can be restored to fertility with chemicals. Science seems to believe that, in any event, it can restore fertility to farmed-out soil by restoring organic matter and allowing soil microflora and insects to repopulate. I wonder. Can a deflowered soil really be brought back to a state of virginity?

At first I thought my confusion was only because of my lack of background in formal soil science training. I just couldn't find any studies that drew conclusions from the myriad of specific comparisons that have been done on the fertility of virgin vs. farmed soils. Nor could I find any surveys or maps showing where virgin soils remain in the United States. Then I got lucky and stumbled upon Ronald Amundson, professor and chair of the Department of Environmental Sciences, Policy and Management at the University of California, Berkeley campus and also a fellow of the Soil Science Society of America. He spoke in a language I could understand (probably because he grew up on a South Dakota farm and has worked farm soil himself). He is as much interested in virgin soils as I am. "I suspect that the biology of virgin soils has more diversity and abundance compared to farmland, but I do not know yet of any microbiologists who have really studied the issue," he said in an email to me back in 2013. "I have been trying to get funding for such a study, but times are tight now." Then he added, most significantly to my way of thinking: "Our soils have been forming for over ten-thousand years. We highly value five-thousand-year-old trees, but think little about completely changing something much older."

I thought of that again when I read a news report on how a company owning a large tract of former forest in California contracted to supply a road builder with nine-hundred thousand cubic feet of roadway fill dirt. It came from a tract of land that had been cleared of redwood forest. Supposing that virgin soil is very valuable as a growing medium, what a tragic waste it was to bury it under a highway bypass.

Wes Jackson, at his Land Institute in Kansas, is keenly aware of how annual cultivation over time can harm the soil, and how reducing soil cultivation through perennial crops also reduces carbon emissions into the atmosphere. The scientists taking part in his experiments are studying just how this happens, and in the process they are learning more about how much perennial cropping can enhance carbon sequestration. Inevitably, this kind of research leads to a question of not only restoring necessary organic matter to the soil, but whether such processes might be able to make modern soils even better than virgin soil.

Just what precisely is virgin soil? Brian Rumsey, an environmental historian writing in the Land Institute's *Land Report* ("Perennial Crops and Climate Change," Summer 2014 issue), makes this interesting comment:

> *Agricultural soil carbon sequestration is not a panacea. Soils cannot sequester infinitely higher levels of carbon. Each has a carbon equilibrium point determined by the plant growth it supports, its physical qualities, and climate. Though some studies have reported higher total sequestration potentials under certain cropping practices, it is wise to suspect that soils cannot sustainably sequester higher levels of organic carbon than they had before they were farmed.*

To be sure of that, we need to have virgin soil with which to compare. And the ability to study food grown on it with food grown on non-virgin soil.

Many of the great books about farming, like Edward Faulkner's *Plowman's Folly,* have argued the possibility that we have "farmed out" our rich soils in ways that modern science can't replenish with accepted soil cultivation practices. Historically, and even prehistorically, humans have used biochar—charcoal—to enrich overused soils. But biochar doesn't seem to have all the agronomic advantages of composts and manures. Organic farmers and gardeners believe that organic practices are the way back to virgin soil, but again, as far as I know, there aren't any really complete and detailed examples of that having been achieved yet.

William Albrecht and his disciples suggest that we have not only farmed out the original organic fertility of the soil but the original nutritional qualities that affect mental health as well as physical health. In other words, the seemingly wild absurdities of modern human conduct might come from eating food grown on farmed-out soil. I might believe that, but history shows that humans were acting out wild absurdities long before the modern era.

I prefer Aldo Leopold's kind of observations in *A Sand County Almanac*, first published in 1949. The whole force of this wonderful book is to argue that neither soil science nor practicing farmers take into account the whole of nature that makes good land and good soil. Riding in a bus through the prairie corn farms of Illinois, Leopold bemoaned the lack of attention or recognition shown to all the causes and effects that go into making a good corn crop. "That the prairie is rich is known to the humblest deer mouse; why the prairie is rich is a question seldom asked. . . ." Looking at the corn growing in the prairie fields and the plants in the fence rows, he groused:

> *No one on the bus sees these relics [wild plants]. A worried farmer, his fertilizer bill projecting from his shirt pocket, looks blankly at the lupines, lespedezas or Baptisias that originally pumped nitrogen out of the prairie air into his black loamy acres. He does not distinguish them from the parvenu quack-grass in which they grow. Were I to ask him why his corn makes a hundred bushels [remember, he was writing this in the first part of the twentieth century], while that of non-prairie states does well to make thirty, he would probably answer that Illinois soil is better.*

But, as Leopold goes on to write at length, the farmer would not know *why* his soil was better.

That virgin soils in the first years of cultivation could have dramatic yields, even by today's standards, is evident from reading old books on farming. Where average yields in the late 1800s (before hybrid corn) were around thirty bushels per acre and up to sixty with good manuring practices, there were yields of over one hundred bushels per acre reported. One of the Leaming strains of open-pollinated corn used in some of the first hybrid corn crosses was yielding 121 bushels per acre, according to A. Richard Crabbe in his book *The Hybrid Corn Makers* (1947). Since open-pollinated corn often lodged badly and produced quite a few sterile stalks, such a yield at that time, achieved without high-velocity chemical fertilizers, would certainly mean an astounding natural fertility in the soil. But so far as I have been able to find, there is no reference ever made to how many years the soil used to grow these yields had been in cultivation—how long ago it had been virgin soil. History says that the Sauk and Fox tribes in northern Illinois, southern Wisconsin, and much of Iowa, were growing about eight hundred acres of corn in 1830, but I can find no details of how much that corn yielded. I doubt that the Native Americans—or for that matter the white settlers who ran them off their land—actually recorded early yields in bushels per acre. But production must have been impressive because it built a great meatpacking industry even before the Civil War and certainly before hybrid corn. In his book mentioned earlier, Crabbe gives this intriguing statistic: "In 1880, our farmers were raising more than thirty-four bushels of corn for every man, woman, and child in the United States."

Forming conclusions about the original fertility of our soils is difficult because there was not as much scrutiny of agronomic records in the days when those soils were first being deflowered. It is possible that I overrate virgin soils. Prince Kropotkin, in his challenging book, *Fields, Factories and Workshops*, originally published in 1907, includes some very complete statistical studies on worldwide farming in the 1880s that indicate, as he pointed out (on pages 75–76) that often crop profits were mistakenly attributed to soil fertility, when in fact other influences were in play. In comparing the profitability of grain grown in Russia and

England from 1860 and 1880, Russian grain realized much more profit because of lower land rents. "The false condition of British rural economy, not the infertility of the soil, is thus the chief cause of the Russian competition," he opined. He then went on to make some intriguing observations about American farming, quoting various statistical sources, plus his own observations:

> The conclusions . . . were fully corroborated by the yearly reports of the American Board of Agriculture and . . . fully confirmed by the subsequent reports of Mr. J. R. Dodge [Farm and Factory, published in 1884]. It appears from these works that the fertility of the American soil had been grossly exaggerated, as the masses of wheat which America sends to Europe from its north-western farms were grown on a soil the natural fertility of which is not higher, and often lower, than the average fertility of the unmanured European soil. . . . If we wish to find a fertile soil in America and crops [of wheat] of from thirty to forty bushels [per acre], we must go to the old Eastern States where the soil is made by man's hand. . . . The same is true with regards to the American supplies of meat. . . . [T]he great mass of live stock which appeared in the census of cattle in the States was not reared in the prairies but in the stables of the farms, in the same way as in Europe; as to the prairies, we find on them only one-eleventh part of the American horned cattle, one-fifth of sheep and one-twenty-first of the pigs.

No doubt Kropotkin had his own axes to grind, always favoring small, intensive farming over large-scale farming, but he does offer something of an argument, or a hope, that perhaps soil can be made better for food production than it was when it was virgin soil. And so much for all my adulation of those hard-driving, hard-riding, western cowboys providing America with its meat.

The notion that our virgin soils possessed an almost infinite and endless fertility is important in cultural history as well as agricultural history. The almost idolatrous faith in the seemingly unlimited fertility of America's Midwestern soils, so well documented by Henry Nash Smith in his *Virgin Land: The American West as Symbol and Myth* (published in 1950), influenced our history in many ways. While our literature and political bombast extolled our Midwest as the Garden of the World, the place where genuine democracy arose because of all those prosperous little farms built on free soil, this led to another interesting idea. Frederick Jackson Turner electrified the scholarly world in an essay he read before the American Historical Association in 1893, titled "The Significance of the Frontier in American History," in which he pointed out that for a century or so there was free land available for settlement for anyone with the gumption to settle it. That fact alone, he tried to argue, resulted in the kind of freedom of opportunity in which democracy, real democracy, could grow and blossom and make America the most powerful and freest country in the world. Turner's "hypothesis" became, for a while, the leading cultural interpretation of American history, but over time it would be partially rejected, if for no other reason than it glorified rural America at the expense of urban America. Couldn't have that. Also it emphasized *free* land as the necessary ingredient, and that idea would always stick in the throat of the money economy, which saw land as a commodity or resource to be turned into cash. The notion that democracy might not be able to exist for long if land were *not* free was way too revolutionary for any practical capitalist or socialist to consider seriously, so Turner's hypothesis over the years receded from the main attention of cultural historians. Only contrary farmers, and not many of them, see the advantages that might occur to society if farmland were removed, at least partially, from the world of commodity buying and selling.

If farmland continues to be a commodity to buy and to sell, as it surely will be, its worth will increasingly depend on how virgin soil is defined and valued. In his most interesting new book, *The Local Yolk* (mentioned in chapter 3, "The Economic Decentralization of

Nearly Everything"), author John Emrich suggests that from an investment point of view, organic farmland will be a much better investment than non-organic because in fact it results in more sustainable food production. And though he doesn't use the word "virgin," his definition of sustainable implies a kind of farming that rejuvenates soil in that direction. Emrich had twenty-five years of investment and corporate finance experience before he decided to concentrate on investment opportunities in sustainable farming—surely another sign that the localized food movement, which he champions, has arrived.

With corn yields now commonly reaching two hundred bushels per acre and sometimes over three hundred (and in fact over five hundred for the first time, in 2014) perhaps deflowered soils are capable of becoming more fertile than virgin soils. Such a suggestion takes the argument into the never-never land of nutritional value. Is three-hundred-bushel-per-acre corn today equal in food value to the one-hundred-bushel corn of yesteryear? Plenty of nutritionists today respond in the negative, but it seems impossible, strictly through scientific analysis, to prove the debate one way or the other. But there is now a concerted effort to take the health value argument in a different direction. The new artisanal farmers are insisting that the soil is the real key to better-*tasting* food, the assumption being that better taste is the mark of better soil. These farmers are going to very complicated methods to enrich soil to achieve that goal. There are examples everywhere. I just ran across one that is particularly to the point: the Lakeview Organic Grain farm in upstate New York. Its best-known artisanal food is flour from emmer, an old form of wheat. The taste of this emmer is highly prized by gourmets and food artists. It is mostly purchased by upscale restaurants. The farm grows lots of other unusual crops, including mustard, kidney beans, cowpeas, Austrian winter peas, barley, oats, rye, and millet. Interestingly, not all of them are grown for market but as part of a very complicated, varying rotation aimed at maintaining fertility, and not only for the purpose of avoiding bugs and plant diseases without toxic chemicals but to *increase the taste of the food.* The owner of the farm says that the unique taste of the emmer *is not about the wheat but the soil.* Taste is highly subjective, of course, but it would be most interesting to compare the taste of his emmer grown in his improved soils and in various virgin soils versus the same crop grown on farmed-out soils.

It is not just fantasy and wishful thinking to assume almost magical possibilities in virgin soil. Recently a World Health Organization report described newly discovered natural clay deposits that have antibacterial properties. Because of so much antibiotic resistance cropping up everywhere, these clays could be a godsend. It appears that ingredients in them—so far discovered mostly near Crater Lake in Oregon—can stop attacks from some antibiotic-resistant bacteria that cause chronic skin infections. Apparently the clays were formed by volcanoes hundreds to thousands of years ago. The volcanic eruptions produced "silica-rich magmas and hypothermal waters" that may have contributed to the antibacterial properties. How's that for magic and mystery?

What virgin soil remains in our heavily farmed areas is mostly in little islands of tree groves that have managed to escape the bulldozer and a few tracts of virgin prairie. Many of these treasure troves of soil are owned by small garden farmers. I almost hate to write about them out of fear that if the marketplace gets a notion, real or imagined, that such soils have a unique fertility not found in even our best farmed soils, there might be a rush to mine them to sell as yet another miracle fertilizer. Or following John Emrich's thinking, cited above, my little grove may be a better investment than stocks or bonds. But I doubt exceedingly that those of us who love our land would ever let it be mined away. Nor would it be practical to clear the tree groves that guard this soil to grow some fantastic corn crop because in short order whatever magical ingredients that might have been in that soil would soon be long gone if planted to corn by modern farming methods.

But there may be ways to use remaining virgin soils for food and fiber production without clearing and cultivating them. Agroforestry has made great strides in this area. In *Plowman's Folly*, mentioned earlier, author Edward Faulkner quotes a passage from Ben Ames Williams's book, *Come Spring*, that just might give an enterprising garden farmer an idea about using little patches of this woodland or prairie soil without decreasing its fertility. He describes an early pioneer farmer planting corn in a clearing where logs and stumps made cultivation impossible, "poking a hole in the ground with a sharpened stick, dropping in two or three kernels, brushing earth into the hole with his foot." Then Faulkner goes on: "I have witnessed the planting of a number of fields of corn in our own time by much the same method. . . . Prodigious crops can be produced by such apparently careless methods in such an environment. Two hundred and fifty bushels per acre are an easily possible yield." He was writing this before 1943, remember.

There are easier and potentially more profitable crops that could be planted on this kind of virgin soil in the partial shade of tree groves. As I said above, giant ragweed abounds in the semi-shade of my woods. The chickens love the seeds. Sheep and cattle love the leaves. Or why not try burdock, which also grows wild in the woods. Have you tasted delicious stir-fried burdock root (gobo) lately? Or how about a cup of burdock root tea? A box of twenty-four tea bags retails for around $7.75 these days.

CHAPTER 18

The Resurrection
of a Really Free Market

S o much pontification balloons up on the horizon every time
the idea of a free market system is mentioned that I have
given up trying to define it. Nothing in life is free. But I
know an almost "free" market when I see one. Imagine that you are
taking a drive into the countryside on a pleasant summer Sunday
afternoon. You pass a farmer sitting out in her yard beside tables
laden with fresh vegetables. Her husband is just at that moment
walking from the field next to the yard with a bushel basket full
of fresh sweet corn. The sign says: organic sweet corn, $4.00 a
dozen. Ah, sweet corn fresh from the stalk. And the price is only a
little above the price at stores. You stop, ascertain that the corn is
yellow—your favorite kind—and of the kernel age that is to your
liking—still very juicy, not waxy yet. You can squeeze a kernel and
the juice pops out. No sign of earworms. (Earworms are easily
removed, of course. But that's another subject.) You fork over your
money and go home to boil your corn. It is the best you have ever
tasted, and you resolve to go back to the same place for more.

You have just taken part in a pure, undiluted, genuine, free
market transaction. But its full import may not have been evident to
you. Note that the four dollars you spent was for food—something
essential to life, not some trinket that will spend but a short time

in your house on its way to the overloaded landfill. The cobs can go on your compost pile. The waste your body expels as defecation is actually very valuable as fertilizer, even if society as a whole has not yet awakened to that fact. The money you spent for the corn will not likely get bundled into an investment fund and gambled away in some banking den of iniquity hundreds of miles away. Your spent money helps the farmer to continue a way of farming that keeps the soil increasing in fertility while it keeps on increasing the good taste of what grows on it. You have, thusly, contributed to the continuance of civilization. The money will work its way through the community, mostly being traded for other goods and services, multiplying its positive economic effect. You have just contributed to the church of everlasting life.

Your corn purchase is not only an example of free marketing but also extremely simple marketing. No expensive advertising was necessary to move the product. No government service was needed either to grow or to market that corn. No heated and air-conditioned grocery store was used. No cadre of middlemen was involved. The corn was not shipped in from far away. The only transportation cost was your car fuel, and you were going to go for a drive that day anyway. No huge orgy of gambling on various boards of trade was necessary to "find a price." The farmer charged what she did because if she tried to charge more, the farm stand down the road would have gotten her business. She did not have to depend on an outside source for her seed, but developed her own open-pollinated heritage variety.

Of course such a simple food transaction can deliver only a fraction of the food needed by the human throng today. But it is the blueprint for farmers' markets, one of the freest commercial market activities yet invented. What that farmer was doing with her corn is very close to what local farmers' markets do for larger populations, only instead of the consumers having to travel to her roadside stand, farmers bring their food closer to where large numbers of people can get to it more conveniently. Farmers' markets are a natural economic outgrowth of urban society and probably ought to be called city markets. They have been around as long as towns and cities have existed and are one of the few economic inventions that have continually resulted in peace and friendship between rural and urban societies. All segments of society mingle at a market: rich and poor, theist and atheist, liberal and conservative, artist and scientist. At a farmers' market, you might catch a Tea Party Republican smiling at an Obama Democrat, both carrying a bag of new potatoes. You might even find, in Baghdad, a smiling Christian buying fresh greens from a smiling Muslim. Farmers' markets are the democratic melting pots of human economics. In them, good food becomes a more effective force for peace than all the great and wise outpourings of philosophy combined.

But it does not happen easily. Free markets are so good at what they can do that both scoundrels and saints are tempted to take advantage of them. In the history of the United States, that vaunted invention of free farmers' markets has not always come easily. In his book *The Farmer Goes to Town* (1948), John Brucato, the founder of the San Francisco Farmers' Market, tells the epic and almost unbelievable story of the struggle he and local farmers went through to get established. In those days, the usual commercial way that food found its way from farm to city was through distributors or middlemen, whom Brucato called "commission men" and whom were exceedingly hostile to the idea of farmers' markets. They contended that so-called "middle men" were often necessary in the complicated procedure of getting food (or any other product) from the maker to the final buyer. This was especially true in earlier days when American farmers were loath to deal directly with the public or with retail stores. In this case they were accustomed to selling their fruits and vegetables to the distributors' headquarters in downtown San Francisco, which in turn sold the produce to the retail stores on commission. When the new Farmers' Market opened, it was met with huge support from the public since the prices were lower and the produce generally fresher, but wholesalers and distributors saw the Market as a threat to their profits, and for four years waged an intense campaign to stop it. All in vain. Twice the matter was put to a vote of the people, and twice the people voted for the Market by a large margin in spite of the big money and big politics pitted against it. The whole battle, in the end, was rather silly because, in reality, the Market not only did not hurt the business of the wholesalers and distributors but actually increased it. Also the wholesalers rarely had the wholehearted support of the retail grocers because the latter were buying directly from the farmers, too, like the public was. Once more the often proven experience in commerce was true again: Just as it is usually good for business to have several fast food outlets next to each other, even though that looks like more competition, so, too, with the wholesaler-retailer complex and the farmers' market complex. They strengthened each other. Let the market run free.

And free, it surely is running. As of August 7, 2014, the last time I did the numbers, statistics showed that in six years the number of farmers' markets had more than doubled, from 4,685 in 2008 to 8,268 in 2014, and the demand continues in all parts of the country. Populous California and New York led, at that time, with 764 and 639 markets, respectively, but Michigan, without a high population, still had a creditable 339. And I'm sure these numbers do not begin to count small, temporary stands along country roads. According to the USDA, all indications are that the number of markets will continue to rise.

But if you are considering starting or joining a farmers' market, know what you are up against. First of all, it is imperative to have a good relationship with the town or city where the market is being established. This means city officials who understand the problems of the market, and marketers who understand the problems of the city. Without mutual cooperation, the market will go nowhere. The more popular the market, the more pressing will be the problems of finding adequate space for it and adequate traffic flow and parking for customers. Invariably public money as well as public cooperation gets involved, and both officials and the tax-paying public need to know that a successful market more than pays its way in direct and indirect increases in city commerce.

When I interview farmers' market managers, it appears that the main ingredient of success is establishing a good working relationship with the local Health Department. It is no good to treat health inspectors as if they were bumbling bureaucrats, even if they are. Farmers' marketers must engage them, not enrage them. If you are obliging and humble, most of them will reconsider stupid rules, especially when the marketer engagingly agrees that most of the rules are not stupid. Regulations can be used to advantage. A good selling point always is to tell customers, as piously as you can, that "we work closely with the Health Department to bring you the safest food in the world." In dealing with inspectors, remember that they don't make the rules, they only enforce them. Don't berate them. And be careful about going over their heads, too, as nothing will irritate an inspector any more than that. More than one farmers' marketer has told me that mutual respect is the key to getting along. Diplomacy is an utter necessity. Don't try to win with an "Occupy the Health Department" offense. Be ingratiating. Sickeningly obsequious. I used the dying lamb approach on milk inspectors when we were milking a hundred cows. It is extremely difficult to milk that many cows and be perfectly, 100 percent, in conformity with all rules of hygiene all the time. So if, on occasion, our bacteria count was too high and the inspector was about to lower the boom, I would hang my head like a penitent murderer on the way to the gallows and grovel at his feet. I am a master at the hangdog expression so essential to survival in the seminary. The inspector always felt pity for me and would give me a chance to get the count down by next inspection.

Being diplomatic with inspectors has led to some real improvements in farmers' marketing regulations. In some states, the Farmers' Market Coalition has gotten special exemptions for "cottage industry" products from regulations that apply to large, commercial operations, like baked goods that marketers make in their own kitchens on a very small scale. The ingredients they use, like their own eggs, may be exempted from the rules commercial bakeries must follow. But it takes the patience of Job to get this kind of legislation enacted. Not only do you have to talk common sense to the food inspection bureaucracy but also convince the moneyed aristocracy of the manufacturing world that being a bit lenient to the little guys will mean more business in the long run for the big guys, too.

Many farmers are at first a little shy, or at least they used to be, about selling their produce. They are hesitant to charge what they know their food really cost them to produce, plus a decent profit. If they have customers from other cultures where bargaining is part of the fun of going to market, they tend to be embarrassed. All they have to do to get over that shyness is to watch the veterans at work. I enjoy going to the West Side Market in Cleveland just to watch the action. Some of the sellers become talented circus barkers unafraid to brag shamelessly about their products, and in truth, the customers sort of enjoy the sales pitches.

Within any group of farmers eager to rent space at the market to sell their wares, ticklish problems can arise. Usually, producers are eager to cooperate with each other for the common good, but as more than one market manager has told me, "there is always one or two who make it necessary to establish set rules for all." Setting prices is always a sticky problem. The law decrees that farmers' market managers can't formally set prices, which is a godsend for the managers when occasionally newcomers with a surplus from a big garden come to market and are willing to sell very cheaply just to get rid of their supply or just for the thrill of being part of the market. The managers can just shrug and point to the law. They can't set prices. It is up to all the marketers to get together and hammer out a policy. "Most often," one manager tells me, "if I go to the price-cutter and talk real nice about how they are making problems for the serious farmers, the newcomers will oblige graciously. The most important ingredient of a successful farmers' market is a manager who knows how to humor people."

Another kind of dumping is more difficult to control. If, towards the end of the marketing day, a farmer sees that he or she is not going to get everything sold, the natural inclination is to lower the prices radically rather than haul the stuff back home. But customers catch on to this tactic in a hurry, and many will simply wait until late in the day to start buying. In such a situation, veteran marketers become afflicted with what looks like the same problem retail stores have with them: fear of being undersold. Suddenly good, old, free enterprise, which they pride themselves on, doesn't look so great.

If you are the market manager, what do you do? Beyond reminding one and all that you can't fix prices, you try to use your powers of persuasion to convince the price-cutter that lowering prices too much is not in anyone's best interest. A manager who is not an astute student of human behavior is probably not going to last very long. "Above all," says Jan Dawson, who with her husband Andy Reinhart operates Jandy's farmers' market (mentioned first in chapter 1, "No Such Thing as '*The* American Farmer'") and who has been very active in the success of the market she helped establish in Bellefontaine, Ohio, "you have to run a very democratic opera-tion. You have to get the growers involved, have lots of meetings where the farmers get together, talk things out, and get the majority consensus on how to proceed." What her market did to discourage dumping late in the day was to draw up a formal agreement among the growers to post an agreed-upon, set price at their stands at the beginning of the day and stick by it until the day is over. So far it's worked quite well. Often marketers can feed leftover produce to their livestock or have arrangements to sell it cheap to other livestock farmers when they have to take it back home.

Farmers' markets don't all follow the same set of rules and regulations. Some require that the farmer produce everything he or she sells. Others allow selling a small amount of produce from another local grower. Some are strictly food markets; some allow homemade crafts. Some have a special day for crafts. The more successful a market becomes, the more it will attract sellers of all sorts of merchandise, and if some restrictions are not in place, a farmers' market will soon look more like a flea market. I was invited to sell books at two different markets, and did, but I considered it a special favor when they were just getting started and did not try to sell a second time. I truly think that restrictions need to be in place to keep the market mostly an outlet for fresh farm food and maybe foods like baked goods processed on the farm with farm-raised ingredients.

The number of products just in the food category and the number of ways to merchandise them are so mind-numbing that a market almost by necessity may have to pick and choose. As of 2015, when I last checked, Kevin Scheuring at Coit Road Farmers' Market in Cleveland was specializing in spices from around the world—over 250 kinds of salt, pepper, and spices. At Local Roots Market and Café in Wooster, Ohio, one of the places I once sold books, Martha Gaffney of Martha's Farm not only was selling her salsa but sometimes entertained her customers with dances to music of her homeland, Ecuador. Many markets now have live music entertainment during market day. At Gordon Square Market in Cleveland, Eric Welles of Skye LaRae's Culinary Services does cooking demonstrations with seasonal foods. Our friends Russ and Beth Miller at the Logan County Farmers' Market in Bellefontaine, Ohio, prepare and sell their own special garlic-flavored sausage—made from their own home-grown garlic and meat—which is a huge favorite among the customers. Jandy's, mentioned first in chapter 1, "No Such Thing as 'The American Farmer'," sells home-grown "cucamelons" (Mexican sour gherkins), the new craze in salad ingredients. At the Gervani Vineyards Farmers' Market in Canton, Ohio, you can do your food shopping and then take a tour of the vineyard in the evening. Some farmers' markets give customers free demonstrations in canning. The Coit Road Farmers' Market (eighty-four years old) holds an annual corn roast, and everyone gets a free, roasted ear of corn. At many farmers' markets, producers will bring along a farm animal or two to liven up the occasion.

Something about a farmers' market encourages geniality. Maybe it is the open air, maybe the thrill of face-to-face encounters with the people who actually produce what they are selling, maybe the satisfaction of buying food so recently harvested from the soil where it was grown. Growers and customers both find themselves standing around talking, sharing experiences, bemoaning the latest world news, making nervous wisecracks about climate change. More and more farmers' markets are finding it worthwhile to put up a tent with chairs and serve free coffee to encourage the feeling of fellowship that pervades the air.

Farmers' markets are so popular now that there is a tendency to move indoors and sell year-round. Local Roots Market and Café, mentioned earlier, is a recent example of how this can be very successful. It calls itself "Ohio's first all-local food shop." Actually, going to a year-round indoor facility is part of the natural, historical progression of successful farmers' markets. The West Side Market in Cleveland is an old and awesome example. Its large building is as much a tourist attraction as the food is a food shopper's attraction. I promise you that, if you have never been to a large food market like this one in operation, your jaw will cramp from sagging open all the while you walk through that place. It is half circus and half town fair, mind-numbing in the variety of foods for sale. Don't go in there hungry or you will come away with an empty purse. It is in fact easy to see how our country fairs got started. First they were farmers' markets, and then human nature took hold and made fiestas out of them.

Chapter 19

✌

Artisanal Food in the New Age of Farming

T he driving force behind the movement from larger to smaller farms is the growing interest in higher-quality "artisanal" foods: fruits and vegetables grown on garden farms, and meat, milk, and eggs from animals grazed on pastures, all of which are located near enough to markets that they can be sold fresh from the farm. Artisanal foods suggest the kind of high nutrient value and high taste that can be achieved in small operations where extremely high soil fertility can be maintained, toxic chemicals mostly dispensed with, and food harvested at precisely the best time and with gentle handwork, to preserve the best taste. Trying to operate a large farm to produce these foods is very difficult—impossible in my opinion. I like to think of artisanal food as the garden farmer's and pasture farmer's parallel to the woodcrafter's handmade furniture. And the same economic situation applies to both in the sense that not everyone appreciates handmade products enough to pay more for them than for factory-manufactured products. The challenge to artisanal food producers is to convince consumers that the higher cost of their food is worth it both in terms of taste and health benefits, and just as importantly, to find ways to reduce that cost enough so that poorer people can afford to buy it, too.

It is folly to try to define good taste, so the producer of artisanal food must keep one eye on what customers think they want and the other eye toward what they might think they want next. The ancient Latin platitude *tot capita, tot sententiae* ("many heads, many opinions") is the only measure that applies here. In this case, listen closely to what the tasters tell you and be prepared to go in another direction or several directions at the same time. If your customers prefer your hamburger over your steak, grind the steak into hamburger for them. If they don't want meat at all, sell them vegetables. If they think your prices are too high, point out to them how many hours you work and then ask them how much money they think you should earn per hour. Sometimes facts need to be faced. Some pricey delicacies might be healthful and delicious but cost a fortune because of rarity or amount of fossil fuel involved. The pious gourmet singing the superior healthfulness of some esoteric seafood does not want to be reminded that his high-priced health food requires burning ten times more fossil fuel energy than a lowly hamburger from a grass-fed cow. If you are a marketer, you must keep yourself delicately balanced between producing practical, low-cost, grass-fed hamburger and high-cost lobster or truffles. (Believe it or not, there are experimental truffle farms starting up in the United States. Check your search engine.)

Humans are very persnickety and fickle when it comes to food (and everything else). How many diets have risen and fallen over the past fifty years? They have all failed, generally speaking, since obesity continues to be on the rise—about a third of the human population right now is overweight. In response to that fact, fad dieting lurches out in all directions. Right now, wheat is considered almost poisonous in some quarters because it has gluten in it. There are some people for whom gluten is rather poisonous, but it has been part of the human diet almost as far back as those apples on that fateful tree in Eden were declared *verboten*. All religions love to forbid some foods and canonize others. Science, or at least pseudo-science, likes to get in on the action, too. Some biblical misanthrope decided that pork chops would send one to perdition just as some scientists only fifty years ago decided that pork chops might cause heart attacks. Today, pork chops are slowly being restored to healthful legitimacy, in science if not religion. The saving grace of science is that it will admit mistakes quicker than religions will.

Humans keep falling for diet fads because they fear death, I guess. Anything that promises to keep one from dying sooner rather than later or from dying at all (as in the Christian bread of life) is worth a try. As an old farmer put it to me one day, echoing Blaise Pascal (although he didn't know it): "I don't believe in all that religious stuff, but I keep going to church just in case I'm wrong."

But faddists prone to jump into every restrictive diet that comes along have spawned another kind of human: the contrary eater who believes in none of them. Food fad atheists. It is absolutely mystifying to them, especially one who is also a contrary farmer, that butter, cream, red meat, bread, and eggs have at one time or another over the last forty years been listed as unhealthy foods and then later been officially resanctified. Artisanal food producers must be aware of all the societal nuances and listen to the contrary eaters as well as the food faddists to serve all of them, including the current plague of diet-crazy locusts. Actually, that's a bad metaphor because locusts are a staple food in some parts of the world and could become so in other parts.

When scientists discovered that there was *some* correlation between *some* foods that *some* people eat with *some* heart disease at *some* times, they built a Tower of Diet Babel. I attended, as a journalist, the conference where scientists first announced to all the world the conclusions of the Framingham Heart Study. Too much fatty foods can cause blocked arteries, the headlines blared. Being at the meeting was a whole lot different than reading about it in the newspapers the next day. At the conference, there were at least as many doctors and scientists present who questioned the idea, or outright disagreed, or at least stressed that more study needed to be done before meat and eggs joined the forbidden fruit in gardens of paradise. But caution rarely makes the headlines.

It has taken over fifty years or so to see fatty foods returned to at least conditional favor. Not that it makes much difference because most people might have given lip service to the High Church of Low Cholesterol, but they kept on downing as many steaks and ice cream cones as they ever did and hoped to make up for their sinfulness by swallowing a lot of statin pills to keep their arteries open, which practice now is also being questioned. We are at a point in diet history where doubt holds the upper hand, and that is usually the beginning of wisdom. As *The Atlantic* magazine said recently (an article in the October 2014 issue, "What Happens When We All Live to 100?" by Gregg Easter-brook): "The Framingham Heart Study, in its 66th year now . . . still struggles with such questions. You should watch your weight, eat more greens and less sugar, exercise regularly, and get ample sleep. But you should do these things because they are common sense—not because there is any definitive proof that they will help you live longer." Artisanal food producers need to keep up with food news like this so they can bring it up in discussions with their customers. But bring it up meekly, tentatively. Just as it is almost always fruitless to argue religion, never argue about taste.

When I think of my food intake over the years, I can only shake my head in wonder that I was lucky enough to stay healthy until my eighties. From my own crazy eating habits from childhood, I should have been dead long ago. I am proof that society today is going overboard on what it calls food safety. During wheat harvest, it was my job as a child to ride on the grain combine (harvester), leveling the wheat as it came from the spout into the bin to make sure none spilled over the sides. I loved to chew on mouthfuls of wheat grains, first making a wad of gum as I chewed and then gumming the wad until it slowly dissolved down my throat. (Hey, maybe there's a new artisanal food—wheat gum.) Obviously, I didn't have celiac disease or I might have keeled over dead in the wheat bin. I knew nothing about that but did worry a little over the dead and dying grasshoppers, stink bugs, and crickets that were mixed in with the grain tumbling into the combine bin. The grasshoppers were especially gross because they exuded a brown liquid from the wounds inflicted on them as they passed through the whirling thresher beaters. It looked like tobacco juice. I tried hard to scoop up handfuls of grain to chew that had not been in contact with this juice or any insect body parts, but who knows? I survived gloriously. And why not? Fried grasshoppers (in butter I would hope) are considered a delicacy by some contrary eaters and they are artisanal food in some cultures.

Another childhood snack was the sweet wax inside wild honey-locust bean pods. The pods, when brown and ripe, would fall from the trees in the autumn, and I would tear them open to get at the caramel-tasting inner flesh. The cows and sheep liked this stuff, too, and ate the whole pods. Some of the pods that I selected for this delicacy had been stepped on by cows and sheep, with their manure sometimes very close by. Again I survived gloriously and am mindful that coffee made from coffee beans that have passed through the bodies of monkeys is also considered artisanal food today—among the rich in Central America anyway.

Anyone who has gone to a boarding school not funded by the very rich has a good grasp of the health benefits of food, or lack thereof. I went to a high school seminary run by Franciscans who had taken the vow of poverty, and our food tasted like it. I have a hunch that our meals were as close to dietary destruction as the worst dreams of the low-cholesterol faithful. Our three main courses we called "bitch" (fried baloney), "shoe leather" (sliced roast beef from cows that must have died of old age) and "hemorrhage" (we never figured out what this reddish, soupy, lumpy stuff was made of). I loved the fried baloney and still do. There is a restaurant in the village of Waldo, Ohio, famous for selling it today (pay attention, artisanal food producers). There was always some kind of gravy or at least some sort of gravy-like liquid for use on the boiled potatoes that were part of almost every meal. When all other foods disappeared down our throats, there was always white bread (horrors) to fill up on. We could get as much of that as we wanted, which for hungry teenagers was a lot. So that the gravy or other juicy stuff would last almost as long as the bread, we would make shallow puddles of it in our plates and then stamp each slice of bread lightly in it before wolfing it down and grabbing for more gluten-glutted dough. Although we complained, we actually ate it all with gusto, disappearing every meal set before us in twenty minutes at the most. We were nearly all skinny and disgustingly healthy as far as I can remember. When we could manage it, we would sneak off into the woods around the seminary and fry slices of Spam we had smuggled in from home. We loved fried Spam.

With that scenario in mind, tappity tap a smart phone forward to 2014, to the high school where my grandson was benefitting from all this nutritious, green vegetable, golden fruit, and gluten-free baked stuff that was available for lunch. I mean no criticism of President Obama's wife, Michelle, whose heart was in the right place, but green vegetables are only really tasty when they are garden fresh and picked when very young and tender, which is rarely possible with commercial machine harvesters that can only handle them when they are more mature and taste something like honey-locust beans without the honey. So while we literally shoved and elbowed our way into the cafeteria to get at the hemorrhage and bitch almost seventy years ago, my grandson packed his lunch for awhile. He says he just can't eat what the school offers. As we contemplate an agricultural world where artisanal foods dominate rather than corn-syrup-saturated factory foods, producers need to think about how they will handle this situation. Is there not a promising market in figuring out how to get really tasty fruits and vegetables to the schools? This is actually happening in some places; plus some schools have gardens where students help grow the vegetables.

Beyond cafeteria days, I have eaten mostly the farmer meals that were put before me, heavy on meat, potatoes, bread and gravy, with lard double-crust pies soaking in cream for dessert. My grandfather said that we did not have to say grace before a meal that did not include meat, potatoes, and gravy. He lived to be ninety-four. Over the last fifty years when I finally had some say over what I ate, I have continued to eat about the same, but along with truly fresh fruits and vegetables picked when still too young by commercial standards. Peas and lima beans especially must be picked rather immature. Corn for me must be still in the pimply stage, squirting out juice when you bite down on the cob. Tomatoes must be vine-ripened. The Kentucky Wonder pole beans must be cooked long and slowly with ham hocks. To begin the summer fresh fruit time, I drown strawberries with cream, followed by cherry pie (lard crust), then raspberries, also drenched in cream. Muskmelons follow, and my favorite way to eat them is with vanilla ice cream. From the last week in July until halfway through September, we eat corn on the cob direct from the garden every day, lathered in butter. Carol bakes bread from gluten-glutted wheat, and often butter-stoked sweet rolls, cakes, and cookies. If I were young enough to get into the artisanal food market, I would not hesitate to preach that to my customers.

A good example of the way artisan farmers should be thinking is Tom Smith, mentioned in chapter 1, "No Such Thing as 'The American Farmer'," who was head chef at one of the highest-rated restaurants in the Columbus, Ohio, area, The Worthington Inn. He was among the earliest chefs to start buying fresh, organic, and artisanal food direct from the farm for the restaurant. He and his wife Abby moved to a little farmstead in our county a few years ago, and Tom regularly took vegetables they grew to the restaurant. He has recently opened a pizza parlor in our little town and grows vegetables such as peppers to use as toppings on his delicious, inventive pizzas. One of his goals is to bring gourmet artisanal food to the beer and hot dog crowd. Yes! That should be one of the goals for all artisanal food producers.

I cheer the Paleo diet that was so popular in recent years—not from dietary convictions but for ulterior reasons. Paleos believe they must get back to a diet more like what people ate in the hunting and gathering age before agriculture. They prefer wild meat but will settle for grass-fed meat—the closest thing to wild that is easily available. As the population of many wild animals skyrockets, artisanal farmers should be talking to the Paleos about the possibilities. There are millions of deer, groundhogs, raccoons, squirrels, rabbits, feral cats, and loose-running dogs that made good, preagricultural food and are overpopulating the landscape. Time to resurrect some of those old recipes for wild meat and convince Paleos into trying them. Venison and raccoon stew cooked slowly in a crock pot can make good eating and maybe good health food. Specialty wild meats from all over the world do well in high-toned big-city restaurants and might be successful everywhere people still know what my Dad often told me: "Young groundhog is fairly good eatin'." We don't need more farm animal factories; we need more Paleo dieters.

When I think about the upsurge in artisanal foods and beverages, I always remember Bob Evans, the fast food legend, who was also a very contrary farmer, although not many people are aware of that. I felt fortunate when he befriended me. When he got out of the army after World War II, he bought an old bulldozer and proceeded to scrape ragged southern Ohio hillsides free of multiflora rose, sowed grass and clover in its place, and turned several hundred acres into a pastoral paradise of permanent grassland. He was one of the very first farmers I knew to see a practical future in pasture farming. He was also one of the first people around to recognize the significance of climate change, but he thought, far from being the end of the world, it would mean that he could graze animals in southern Ohio all year round, which he eventually did.

I became acquainted with him in my usual stumbling, bumbling way. I had written a sharply critical article about the College of Agriculture at Ohio State University in *Ohio* magazine, for which I was catching hell and damnation from all sides. One of the ag economics professors who seemed so civilized in the classroom went so far as to inform me, on the phone, that I was a son of a bitch. (We later became almost friends.) At the lowest point in my ordeal, I got a phone call. On Christmas Eve, no less. The gravelly voice on the other end announced: "This is Bob Evans."

Silence.

Could it be *the* Bob Evans? Was he going to chew me out, too? After all he ran a rather big agribusiness operation.

"I just want to tell you that you got it right in that article about Ohio State. I've been trying to tell those people the same thing for years now."

We were friends ever after, fat-laden sausage sandwiches and all.

His contention, and mine, was that the university was only serving the large-scale corn, soybean, and animal factory agribusiness network and was ignoring young farmers looking at alternative kinds of food production. (Since then, of course, the professors have caught up with the times and offer all kinds of programs for pasture farmers.) If only the largest farms were being encouraged, we argued, that would inevitably mean fewer landowners, which would eventually mean the death of democracy, in our opinion. Bob was even more feisty toward the College of Agriculture than I was. He wanted Ohio State to come to the aid of pasture-based farming. "Those big corn farmers don't need any help," he would say. He had little use for either traditional or industrial husbandry, since both seemed to ignore the possibilities of pasture farming, which needed no high-cost technology of any kind. He was convinced, for example, that silos were unnecessary. He would mutter to me: "They oughta dynamite them all."

I loved to go into one of his restaurants with him. He would grab a menu, run his finger down the list of foods and instruct me on what to order. "Don't get that. Garbage." His finger would continue to roam. "Get that. It's pretty good." He believed the sausage he was famous for was no longer of the high taste quality he had brought to it in earlier years when he was hawking it to individual restaurants and grocery stores in his area along the Ohio River. He had, by the time I knew him, been ousted from direct active leadership of his own company. The kind of pioneering drive and the extremely hard work it took to turn a home sausage-making enterprise into a huge fast food chain required a kind of spirit that just didn't mesh with the corporate way of keeping a big business running smoothly. I don't think it bothered him much because in his later years he became so involved in experimenting with pasture-based farming that he was glad not to be running the company. He was a contrary farmer, his soul the very antithesis of big corporate business.

The marrow of his philosophy, although he did not exactly practice it himself, at least in the restaurant business, was to advance a kind of small-scale, low-cost agriculture (that is, grass farming) that many families could follow as a way of life, not just the moneyed few. Put your money into livestock that biologically reproduce themselves, he liked to say, not into costly machines that only factories could reproduce. "Tractors don't have babies," was another of his favorite remarks. This philosophy was also behind his entry into the restaurant business. He wanted a place to eat that most people could afford. It may be hard to imagine, but he actually started out very much in the way of the local food movement today. He and his parents raised hogs the traditional way and hawked pork sausage personally from restaurant to restaurant. To him, his food was as artisanal as any being sold today, and who's to say it wasn't? People loved it, and it was priced so that nearly everyone could afford it. So what if the diet religions frowned? They eventually came around.

It's fun to eat truffles and caviar and antelope jerky occasionally if you've got the money, just like it's fun to buy an expensive English hoe, even though a cheaper one will do the job. It is great to be able to pay five dollars for a dozen organic, blue Aracauna eggs or a box of flour made of dried, roasted crickets. But some place along the line, there must be attention paid to producing wholesome tasty foods that are also affordable for most people and require less carbon emissions to bring to market, not more. Even many of the more esoteric new or rare vegetables and fruits can be grown without high cost. Cucamelons (real name: Mexican sour gherkin) can be grown in the North in hoophouses and have become very popular at the Bellefontaine, Ohio, Logan County Farmers' Market in recent years.

One of the wildest new specialty foods is cricket flour. For reasons hard to figure, since we are not exactly short on food in this country, an interest in eating bugs seems to be on the upswing. Pay attention to that. Cricket farming made front-page news in a recent *Farm and Dairy* magazine issue, and this is not a publication that usually goes far out on a limb about far-out farming. There is also a budding interest in growing hops, necessary in brewing beer. The proliferation of local breweries, which have to buy hops from Oregon or Washington, has renewed interest in growing the plant closer to home. Hop farms were once widespread—grown even here in Ohio in the early 1900s on my cousin's farm. When the brewing industry consolidated, so did hop farms. Now the worm has turned again. As beer brewing decentralizes, so does hop farming. Ohio State has been experimenting with this crop at its research center in Wooster for several years now, and a few growers are tentatively beginning to stake out hop fields. Craft beer is also spurring more acres devoted to growing barley for malting in local areas close to the breweries.

The controversy over fats and cholesterol has contrarily opened new farmers' markets. During the scare about cholesterol, Jersey and Guernsey cows, known for milk high in butterfat, declined in numbers and Holsteins, with less fat in their milk, increased. (A neighbor who milked Jerseys told me once that he kept a Holstein in his herd in case he had to put out a fire.) Then, slowly, the attack against saturated fats subsided to the point where books singing the praises of Jerseys and Guernseys found their way into print. People started looking into dairy products with a more discerning eye. Consumers discovered what dairy producers have always known: milk is not a generic product but comes in many versions with many tastes. Jersey milk tastes different than Holstein milk. Cow milk tastes different than goat, horse, or sheep milk. When we went from milking by hand and cooling the milk in tubs of well water to machine milking, with the milk flowing directly from the cow through a pipeline into a cooling tank where its temperature was lowered rapidly, the taste improved markedly. Cows out on fresh green grass after a winter on hay and grain give milk with a decidedly different flavor that takes some getting used to for most consumers. Milk from cows eating mostly corn silage tastes different than that from cows eating only hay and grain. Homogenized milk tastes different than unhomogenized; pasteurized tastes different than raw. Fresh from the cow, milk tastes different than after it is cooled. Even the kind of plants in the pasture can change the taste—wild onions, for instance, make the milk taste sour. Milk connoisseurs discuss milks the way wine connoisseurs compare wines.

All of this opens up a land of opportunity for artisanal farmers, just as the wine business did for grape growers. Heavy cream still is not available in regular supermarkets except as homogenized, pasteurized whipping cream, so local farmers with their eye on the market cater to new demands by adding Jerseys and Guernseys back into their herds and offering customers their own version of heavy cream. Some customers want milk with the cream still on top, so that's what local dairies provide them. Others decide that they want unpasteurized milk. This takes some doing since unpasteurized milk in many areas is as illegal as bootleg whiskey. But where there's a will, there's a way, and raw milk is now legal in many places if you follow all the rules. Some innovative dairies have found success in simply going back to the older, slow-heat way of pasteurizing milk because customers prefer its taste to flash-pasteurized milk. When I heard milk aficionados here in Ohio arguing about which slow-pasteurized brand had better taste, Hartzler's or Snowville's, I knew we were in a new agricultural era. And now there is another development, or new interest in an old development. It appears that some milk contains mostly A1 protein and some contains A2. The A2 kind, more often found in Guernsey and Jersey milk, is being touted as healthier, especially for people with lactose intolerance. The matter is highly controversial at this point, and science hasn't weighed in on the pros and cons. But again, the controversy is good news for artisanal farmers because it will sharpen consumer curiosity about milk and no doubt mean new products to develop and sell.

We have already seen the yogurt market blossom. I expect that some enterprising artisanal milk producer will take a closer look at cottage cheese, if some haven't already. Our homemade cottage cheese from our own cows always tasted much better to me than anything in stores. And of course, there's butter. Organic Valley's new butters, almost entirely from cows on grazing routines, is getting lots of praise.

The egg market is an even better example. All the controversy over cholesterol prompted consumers to take a closer look at eggs. When the storm clouds cleared, they had learned enough to know that they wanted fresher, locally produced eggs. I remember so well when we first sold eggs and a neighbor called, wondering if there was something wrong with ours. She said the yolks had an unusually rich orange color and stood up plump in the skillet. We explained to her that eggs were supposed to look like that and how our hens had a varying diet from grazing all over the lawn and woods. She called back later and agreed that yes, indeedy, the eggs did taste better. This conversation undoubtedly has happened thousands upon thousands of times in local areas all over the nation, and as a result backyard hens are all the rage today, and farmers and farm suppliers are learning to cater to the needs of backyarders—supplying feed, bedding, waterers, nest boxes, and advice. Who would ever have dreamed that today there would be quite a few successful businesses serving as consultants to backyard food production? One of them, John Emrich, has even written a book about it (*The Local Yolk*, mentioned first in chapter 3, "The Economic Decentralization of Nearly Everything").

The same turnaround is occurring with meat. Today, whether meat is good or bad depends on whom you are talking to, but the argument itself has been good for everyone. Out of it are coming revelations about how industrial meat is being produced and the differences between animals kept in enclosed areas with heavy diets of corn and those allowed to get most of their food by grazing open pastures. Consumer interest in grass-fed meat is increasing faster than farmers can supply it.

The food police decided, even before the Paleo diet came along, that modern grains were bad for one's health, especially modern wheat. The controversy has gotten people to look closer at grains in general, just like they did with meat and milk. They become better informed and more sophisticated about grains. In response, demand is opening up for artisanal foods made with special kinds of grain, some gluten-free for those who need it, and some full of gluten for those who want it. All kinds of old and esoteric grains are being grown now. For example, emmer, an ancient form of wheat, is becoming popular for bread dough. A few years ago, very few people knew that emmer even existed.

CHAPTER 20

⟵⟶

Why Fake Steak
Won't Ever Rule
the Meat Market

Several years ago, science achieved a breakthrough on the artificial meat frontier: a little round slab of something that looked sort of like a burger, smelled sort of like a burger, and according to the official testers, tasted sort of like a burger but needed some seasoning. It was grown in a laboratory from muscle tissue extracted from an unborn calf. Conjuring this miracle into existence cost something a bit over $300,000, so I read. More recently, as experimentation continues, one maker of what is now called "cultured" meat, says it has got the cost of its lab product, brand name Memphis Meats, down to about $18,000 a pound. It grows the tissue in what it refers to as a bioreactor tank.

Reminds me of the hue and cry over fake butter many years ago. Learned prophets had margarine sweeping butter off the table. Quite the opposite has occurred, as butter regains its hold on the taste buds of humans despite all that propaganda about cholesterol. Then came the many attempts to replace cow's milk with soy milk, almond milk, and other concoctions. That effort is still ongoing, but cow's milk rules.

It seems to me that the farm unrevolution ought to make a statement pro or con on this amazing fake meat achievement. Science has dreamed of producing meat in the laboratory for at least half a century, and even more seriously since PETA (People for the Ethical Treatment of Animals) offered a million-dollar prize in 2008 to the first company that could grow chicken breast in the lab before 2012. Now the dream has been shown to be possible. The pioneers in this noble endeavor figure it will take only another decade or so to get the price down to affordable levels.

I don't know how to feed muscle tissue in a bioreactor tank, but I know a whole lot about feeding animals on a farm, and I am convinced lab meat will never come close to taking over the market. High costs should be expected in the developmental stages of any new product, and I shouldn't make fun of that. Artificial meat would have some advantages, at least over factory meat from real animals. But it seems to me that the people involved (investment money is pouring in from Silicon Valley) might want to hear what "get small and stay in" farmers with several centuries of experience behind them have to say before millions more are spent trying to create a Porterhouse steak in a bioreactor tank. Memphis Meats says it hopes to have its shamburger in the marketplace in three or four years, and a full line of meat products including filets in about twenty years.

I am sure that an affordable fake steak, even if it goes to market, will never ever replace the real thing. I will try to explain to well-intentioned people why I say this, without sounding any more arrogant than I have to. I even wonder whether it is correct to say that the people who flatly state livestock are one of our biggest threats to the environment are "well-intentioned," because if they were, they would take the time to get to know cows better, even if bovines do burp and fart in an unseemly manner. But arguing with them is like trying to convince my grandsons that broccoli tastes good. Or like trying to sell a red tractor to a guy who has never owned anything but green ones.

First of all, those people who believe that commercial livestock should be removed from the face of the Earth seem to assume that without farm animals, the Earth would remain devoid of other animals that breath, fart, and defecate as much as cows do. Some of the claims of the anti-livestock brigade in this regard (you can read about them on Wikipedia under "cultured meat") would be too outlandish to deserve attention in any society familiar with how nature works, but unfortunately we do not presently live in that kind of society. "Conventional farming . . . kills ten wildlife animals per hectare each year," claims one defender of a cow-less environment. "Converting . . . 10 acres of farmland from its man-made condition back into either pristine wilderness or grasslands would save approximately 40 [wild] animals. . . ." Do people who make remarks like that not know that the wild animals they want to "save" expel carbon just like livestock do? Even ants emit an awful lot of CO_2, so I have read.

I have no idea how the claim that farming kills ten wild animals per hectare was arrived at, but it will send farmers in my neighborhood into howls of laughter. Deer, raccoons, squirrels, groundhogs, rabbits, starlings, English sparrows, moles, opossums, coyotes, feral dogs, feral cats, foxes, Canada geese, mink, and chipmunks are overrunning us here in Ohio. They shit, too, as anyone knows who has observed what Canada geese can do to a golf course. Wild turkeys and black bears are growing in numbers and becoming nuisances, at least for modern-day mankind, and they are champion defecators. The Ohio Department of Natural Resources was so proud of itself when it reintroduced beaver to our state a few years ago, and now, red-faced, it has been forced to trap and remove them from some areas because of the destruction they are causing.

The whole notion that there are two kingdoms in the countryside, one wild and one domesticated, should not be emulated, even if it were true. Joel Salatin, certainly a five-star general in any ranking of new-age farmers, bemoans the prevalent social attitude that wants to separate wildlife and farming into two separate areas of biological activity. In an article, "Order vs. Wildness" in *Acres U.S.A.* (December 2014), he writes:

> *I think our culture suffers from the perception that ordered farming, or ordered landscapes, must inherently militate against wildness. . . . The idea, perpetrated by Thoreau, that farming order and wildness were mutually exclusive and therefore required segregated and designated areas allows landscape managers to be lazy about wildness. . . . I'd like to see more creativity, more visceral expressions of commercial farming order not only co-existing with wild systems, but actually enhancing them.*

But even where that wise union is not specifically intended, it happens anyway. Grain farming has turned the countryside into almost a paradise for many species of wildlife. Those endless acres of corn and soybeans make an almost unlimited dining table for millions of wild birds and animals. Areas once farmed but no longer conducive to profitable corn and soybean production because of high cost are filling up with brush and trees—excellent cover for wild animals when they are not feasting on crops, like for instance, my open-pollinated corn, which they prefer to new hybrids. Or when they are not feasting on millions of gardens, also like, for instance, mine. The number of farm ponds—like for instance, mine again—has grown exponentially over the years, and along with thousands of ponds created along highways when road-makers need fill dirt, they provide havens for increasing numbers of fish, ducks, and other aquatic animals to increase and multiply. One reason eagles are increasing in number in my state (Ohio) is all this new water. Eagles prefer fish in their diets. But they will eat dead sheep, too, not to mention live lambs. Herons are increasing and giving aquaculture farmers fits. A farmer in my neighborhood tried to start a business raising koi and other fish for stocking backyard ponds but quit, mostly because of raiding kingfishers. Yes, kingfishers! Our daughter and son-in-law, who live in suburban surroundings, have a difficult time keeping fish in their garden pond because of wildlife predation, especially blue herons. The last time I was visiting there, a coyote sauntered across their yard during one of the few intermissions when the deer had vacated the landscape.

Fifty years ago there were no deer in our very rural county, but lots of cows. Today most of the cows are gone as farmers here gave up livestock and turned to grain farming exclusively. Without grazing livestock around, herds of deer replaced them, hiding out in the woodlands and brush and then, when the coast is clear, roaming the grain fields for food. In fact, the coast doesn't really have to be clear. You can stop along any number of country roads on a fall evening after harvest and count thirty or forty of them at a time dashing across the cornstalks. They are causing thousands of car accidents and death to both humans and deer. They have gone to town and are crashing through church windows and storefronts (honest) and are leaping to their deaths from parking garages. They fart, too. Maybe if they start breaking through the windows of the *in vitro* meat laboratories, the scientists will get the message. Nature can make meat cheaper than they can.

Nature abhors a vacuum. If we return the Great Plains to its pristine prairie, the deer and the antelope will play there again, along with wild horses, prairie dogs, coyotes, wolves, endless clouds of birds, and eventually thundering herds of buffalo, all of them defecating up a storm of manure and CO_2 at least as measurable as what livestock do now. And, yes, that could be good because it would definitely mean meat cheaper than anything that can be raised in a bioreactor tank.

As the Union of Concerned Scientists has said, the energy and fossil fuel consumption required to make lab meat might alone be more environmentally costly than the whole business of producing food from the land, especially, I'd add, if we produce it in an ecologically sane way. "Well-meaning" anti-livestock people should be supporting those of us in farming who think that grazing animals on grass and clover to produce meat and milk is much more environmentally sound than going to the horrible expense of annual cultivation of hundreds of thousands of acres of grain and then drying, storing, and hauling the grain to livestock in confinement. With pasture-raised animals, we could not only produce meat cheaper and with less carbon pollution than anything out of some Aladdin's bioreactor in the scientific laboratory, but cheaper than with annual grains. The grass and clover go directly into making meat and milk without soil erosion and without nearly so many farting tractors needed to produce all that annual grain. At the same time, the grassland sequesters a significant amount of the carbon that the belching, farting livestock produce.

When "well-meaning" people today deplore all the pain and suffering involved in raising and slaughtering livestock, they seem to ignore the fact that even more grisly pain and slaughter occur in the wild. Lions and tigers pull down zebras and wildebeests and eat away at them even while they are still alive. Wolves and wolverines chase a deer to exhaustion and then tear it apart bite by bite. Hawks rip squealing rabbits apart with their beaks and talons. Rarely in the wild does anything die of old age.

Education has failed in this regard. We no longer teach children the reality that they learned just by living on a farm a few generations ago. Nature is one huge dining table around which all life sits, eating while being eaten.

The part of our society that is lapsing into livestock phobia ignores the millions of pet cats and dogs that now live as wastefully, in terms of unnecessary and artificially supplied energy, as humans. Cats and dogs expel and defecate, too, and most of their manure is going into landfills at another enormous expense. There are at the latest count seven to eight million cats and about the same number of dogs living the carbon-polluting life of humans. To make things worse, humans irresponsibly turn unwanted cats and dogs loose in the countryside to increase and multiply and kill millions (yes, millions, the statistics say) of beneficial songbirds. If we would eat these cats and dogs, as people do in other parts of the world, we might help the problem, but most Americans are as horrified of that as of eating horses.

Horses make an excellent example of how the anti-livestock people are living in an unreal world. Because of the misplaced devotion of some horse lovers, many of whom have never raised a horse, the Bureau of Land Management is not allowed to cull the wild herds that are increasing and multiplying. Horse slaughter plants have been closed in this country. Just as irresponsible people abandon cats and dogs, so they abandon horses they no longer want and have no way to dispose of cheaply. Right now, they must ship the horses to Mexico to be slaughtered. Some seventy-five thousand horses now roam the "pristine wilderness" in the West, and to keep them from starving, the government is corralling and feeding them at what by all accounts is an exorbitant amount of money. I read that the cost is at least $2,400 each per year. Multiply that by seventy-five thousand and you see why this approach is just not going to work. We would be better off eating some of those horses or turning the surplus into pet food and fertilizer.

I wonder if the cultured meat champions have figured out just how awesome an amount of synthetic meat they would have to produce to take the place of the natural meat now being consumed. They seem fixated on cows, but what about the millions of goats that in many parts of the world supply relatively cheap meat and dairy products that semi-nomadic people depend on to keep from starving to death? Then start counting the number of pigs and chickens worldwide.

Also the anti-livestock coalition seems unaware of the byproducts made from animal parts not used for human consumption. Pet food. Leather. Glue. Fertilizer. Soap. Tallow, from which fatty acids are manufactured and then used in pharmaceuticals, petroleum products, plastic products, rubber products, inks, paints, and varnishes. The cost of replacing these byproducts with products from vegetative sources, even if possible, would surely be astronomical.

Personally, I think the argument about whether cultured meat can ever compete with real meat in the marketplace has already been made and won by the latter. It's called Spam.

CHAPTER 21

The Homebodies

G arden farmers are not regular travelers. They can't afford much of it, for one thing. But that's not a problem because they like staying at home. In fact, that is one of the built-in profit-makers of farming—all the money one can save because timely crop plantings and harvestings plus taking care of animals keep the farmer at home most of the time. But most of us in this kind of world don't mind at all. It makes a great excuse when you are invited to something you really don't want to go to anyway.

"I would like to accept your offer to give a speech, but that will be during lambing time when I must be here 24/7."

"I'm sorry to have to leave now that the party is really taking off. Have to milk the cows, you know."

"A week in Florida at your place would just be so wonderful, but I can't find anybody to take care of our livestock."

"Can't go anyplace right now. I've worked all year on these strawberries, and I am going to stay here during harvest season and eat every one of them."

Spring means staying home to get all the planting done at just the right time, and you never know when rain will cause a delay, making planting the next time the soil dries out extremely critical. Then harvest starts with peas, and if vegetables are not picked at just the perfect time, why do it at all? That leaves winter for travel when ice and snow looms overhead. Once you get caught in an airport or on a highway for a long period of time, you are cured of any desire to go far afield again.

In these times of relentless restlessness, it is sort of strange, I guess, that my siblings and I, all nine of us, either never left the environs of our home farm or came back to it. Four of my sisters and my brother live right on the home place, and have for many years. Two other sisters live about ten miles away, and I live two miles away. Another sister, now deceased, lived her whole life here, too. I was the only one to move away for a long period of time, and I came back forty years ago, as soon as I could manage. When we are together, we agree that we all prefer to live "far from the madding crowd." But that merely begs the question. Why?

One sister, the one we call Berny, short for Bernadette, is still living in the house where she was born some seventy years ago. When I reminded her that our first cousin David, a purebred contrary farmer a few miles away, has her beat, having lived in the house where he was born for eighty-eight years now, she arched her left eyebrow and replied: "But does he still sleep in the same room where he was born?"

David smiled when I asked him. He was not sure. He did know that no one has lived in his old farm home any longer than he has. He spent three years in the army but all the rest of his life he has lived there, farming his place. Retired, he has resigned himself to watching a neighbor farm his land while he looks out yearningly over the fields, remembering the only life he ever wanted and did so thoroughly enjoy. He still complains every time he has to go on a trip that keeps him away more than a day. Often he tells me that he would rather sit on his porch and look out over his fields than look out over any ocean in the world.

Berny has lived on our home farm all her life except for about three years when she was first married, during which time she moved hardly three miles away. She still actively farms with her husband, Brad, raising sheep, chickens, and a few beef steers, plus a garden so perfectly neat, weedless, and lush that it would make

Luther Burbank jealous. I am relieved that I do not live next door like three sisters and a brother do, because there's no way I could keep my place as neat as they all do. They all own parts of the original farm. Sister Marilyn and her husband, Dennis, and Brad and Berny own the center part of it, a manicured pasture where they have grazed sheep as did Dad, and before him Grandfather Rall, and before him pioneer rancher R. N. Taylor. It looks more like a golf course than a pasture, and I will pay you five bucks for any noxious weed you can ever find going to seed on it. That's including the land right along the creek that winds through it, land that only hoes, not mowers, can reach. Passersby sometimes stop alongside the road just to admire the scene because in our county of corn, soybeans, and weeds that not even Roundup will control anymore, there is nothing quite like it. Driving over the brow of the hill where the curving creek, hillside pasture, and grazing sheep come into view is like driving into 1940, only back then Grandfather Rall never could keep the Canada thistles under control like his offspring do now.

We all grew up on that pasture. It was our playground, as it has been for our children and grandchildren. For reasons none of us can explain very well, we never really wanted to live anywhere else. We like to quote the lines in Wendell Berry's poem "On the Hill Late at Night":

> . . . I am wholly willing to be here
> between the bright silent thousands of stars
> and the life of the grass pouring out of the ground.
> The hill has grown to me like a foot.
> Until I lift the earth I cannot move.

We spend a lot of time trying to explain to each other why we are so attached to our home grounds. I worry that I make too much of it, but it seems unusual enough to examine closely since quite a few other contrary farmer families feel like we do. At the root of it, what keeps us planted on this land is at the heart of understanding the soul of true farmers. All of us do take trips, but rarely, and then usually to please someone else. One sister and her husband have a cabin in Canada where they vacation. Another sister has been to Europe with her husband. Another travels regularly to the West to visit a son and his family. And certainly, I have had to travel, however unwillingly, as part of being a journalist. Nor do we share a common philosophy other than loving to be at home. I think it is simplistic to try and label any real person as "liberal" or "conservative," especially regarding us. Although we were raised as Catholics, only one belongs to an organized church today. Three of us farm commercially, but only one full-time. All our spouses grew up on working farms except one, and he was born and raised in the nearby village and has had one foot in farming most of his life. One brother-in-law was a plant manager before retirement and, then as now, spends most of his spare time like the rest of us, tending to his farm pond, orchard, gardens, and woodland. All of us keep chickens. Another brother-in-law is a retired realtor and professional photographer. Another, now deceased, was an accountant, and yet another was a rural mailman before retirement. Sister Rosy was a teacher before becoming a full-time mother. Sister Teresa worked in a retirement home and children's daycare center. Sister Marilyn worked as a secretary for a doctor before marriage. Kathleen, now passed away, worked in a factory.

Our homebody genes may have been inherited from our mother, whose maiden name was Rall. When I left home for faraway places, she pointedly told me not to expect her to visit much. It was not lack of love that made her say that. She just thought she had to be at home to make sure all went well with the livestock and gardens and the children still at home. In her whole life, she never saw an ocean or a mountain or the innards of a big city, but read voraciously about the whole world. Once, when I was in boarding school, she and Dad travelled the five hours necessary to see me. They stayed two hours. The farm parents of a Rall cousin who was in boarding school with me came once and stayed three hours. The parents of another cousin never came at all. My cousin Adrian, who lived across the fields from our home place, had a daughter and family who had moved to New York State. She constantly tried to talk him into coming there for a visit. He resisted as long as he could. He told me that he could just not bring himself to leave his livestock in anyone else's care.

Great-grandfather Rall and his brother came here from Germany (so much for being homebodies), walking the final stretch from Columbus, Ohio, leading two cows, so family legend has it. They and their four sons gained enough land over the years to provide a 160-acre farm for any member of the third generation (my mother's) who wanted one. The whole neighborhood became almost entirely little Rall farms. When I was a child, it was possible to walk all the way to town, four and a half miles away, almost entirely on Rall land. Miraculously, most of that land is still owned by members of the family, and except for a few new houses built on the old farms and the disappearance of some barns and fences of the earlier era, the look of the land has hardly changed. I did not realize that there was anything unusual about our situation, or that the era of small family farms was indeed passing swiftly right before my eyes, until it was mostly gone. But, in a strange sort of way, we stopped it from happening completely on our home grounds.

Grandpaw Rall built the barn on our home place about 1920. My father, brother, and I tore it down in 1958. The new barns we built, trying to keep on farming by "getting big," came down about 1980. Brad and Berny replaced some of them with a smaller barn and pursued a part-time farming venture with sheep instead of the full-time dairying that the family had previously engaged in. In only about sixty years, the wonderful agrarian era of small, stable, self-sufficient family farms bloomed and died before our eyes. Almost. My brother, Giles, bought much of the southern half of the farm when Dad passed away. Sister Gerry and her husband bought the biggest chunk of the northern part. Marilyn, Jenny, Rosy, and Berny and husbands bought smaller chunks in the middle. Carol and I purchased old Rall land two miles away for our new home. Then several of us bought the farm adjoining the original home place about 1990. It had been absentee-owned but farmed by Ralls nearly forever. Finally, Berny and Brad bought parts of yet another adjoining Rall farm in 2009. In a sort of surreal way, we kept alive the old Rall family farm kingdom in miniature.

By trying to preserve the agrarian culture of our childhood, I think we are in a way unconsciously fighting time, fighting death. Overcoming physical death is impossible of course, but cultural death can to some extent be evaded if not avoided. Trying to preserve the good parts of our past contrary farmer way of life is what stirs our souls. The "good old days" must pass, but not necessarily for us. Not yet. Not . . . Quite . . . Yet.

Berny, the poet in the family, wonders, not altogether in jest, if our love of home comes from an accidental but right alignment of the stars. "It has to have something to do with the particular decades there on the land where we were born and grew up. We were so *rural*. All we had and all we did involved and revolved around the land and nature. We were such a *group*. We had tremendous solidarity." Then she smiles and suggests, "Maybe it is simply because we all actually like one another." She shared a poem with me that I think could only have been written by someone who has, by choice, lived in the same place her entire life.

A PLACE NO VOICE REACHES
(FOR MY MOTHER)

It is the core of summer. All day long the wind
wastes its breath; no one can face
the thermometer. After dark we'll count
crickets' trills to get the degrees.

In a field the ka-thunk, ka-thunk of a baler
making hay. We used to have lemonade squeezed fresh
in a kettle. The men had beer. The sweetest
sound was the quiet after shutting down.

What is left of you in this place has been turned
under, grown over. I must have worked the ground
a hundred times wanting to free you. Every year more
comes into view. Your voice, lodged in the maples,
wafts down with the merest nuance of the little leaves.

Here in the kitchen where you wintered out the cold,
knew what hour the sun entered the fourth pane to settle,
as now, a few inches from my hand,
I hold the last warmth of you in my skin.

Teresa, one of the two rather committed artists in the family, thinks that a good home life is a key. "If you have loved your childhood, you will find a way back to it as an adult." She left the home environment for a couple of years as "an experiment" to study art in the big city. "The anonymity of the city seemed cool, but that got old in a hurry," she says. "It was so nice to get back here where I had a background, family history, roots, repeating in adult life what I had found to be good in youth." She likes to quote from Michael Paternini's book *The Telling Room*: ". . . I wanted to live . . . where all the generations of my family had once resided, where I might take daily strength in them, and where I'd live a life antlered by meaning and mysticism."

Jenny, the other artist in the family, thinks that, in addition to the comfort of having a cultural identity, the happiness of staying home is a way to feel in control of her life. "I feel so much more secure at home than on the road." Like poet Berny, both painters—Teresa and Jenny—stress that they believe good art can only come from deep familiarity with one's place—another reason why they are happy to stay at home and paint the familiar world around them. "I don't mind being alone," says Jenny. "I don't like loneliness but enjoy aloneness. I think there is a psychological aspect to my distaste for going places that is perhaps unhealthy and easily diagnosed. Even so, I enjoy going out on an occasional jaunt not too far away. Jim [her deceased husband] and I used to do some traveling and had good times, but somehow my uneasiness never left me until we turned around and started home."

Gerry and her husband Emery, who farm about a thousand acres on the other side of our town, moved, when first married, to land in North Dakota that was in Emery's family. That lasted four months. "We had a nice little farm and had lots of fun there, but there wasn't one single thing around that I knew any history about," she says. "Every farm, road, and store was new to me, just a surface without a history. I couldn't find a place where I belonged. I literally didn't belong there." She pauses, growing more pensive. "I've always had a strong sense of family. Even in high school, I sometimes chose to stay home rather than go to social functions. I was not shy or introspective, just liked being at home. In addition, I often disagreed, then and now, with the popular view of things. That sort of kept me from socializing a lot. Still does."

Rosy and I are the writers in the family. For twenty-five years, she wrote a newspaper column called "On Home Ground." Dad paid her to work for him on the farm during summers home from college, and she taught school for awhile. "In college, I thought about becoming a doctor or something like that. I figured I was smart enough to do about anything I set my mind on. Both Otto [her husband] and I earned teaching degrees, and we could have lived about anywhere. But we both wanted a country lifestyle here in our native community."

She makes a distinction between a homebody and someone who decides to live in one's home territory. "I was eager to carve out my place in the legacy I thought I had been given and was a part of. Being a homebody, a stay-at-home, is something else. To be a homebody successfully, one must know oneself through and through and be completely comfortable *and content* with that knowledge. To prefer to be a non-gadabout, you have to be past the stage in life when you are searching for fun, excitement, new experiences, that sort of thing, for the very practical reason that often your only companion is going to be yourself. Married couples who are both homebodies have each other, but the contentment issue there is just the same."

Rosy thinks of herself as being both a homebody and having a deep yearning to stay closely connected to familiar, elemental, homey things—not wondering what is beyond the horizon. "You get closer to the truth the more deeply you can pursue the familiar world around you. When I say we are peasants, I am not really joking."

She explains herself in one of her columns titled "The 'And Yet' Factor" in which she reacts to a reader who thinks of her as "just a garden writer" because often her columns were about gardening. She points out in the column that she does travel some, does socialize, has many interests beyond her garden, that she is well aware that there is "life beyond the bean rows." Then she writes in her trademark conversational way:

And yet . . . and yet. It is almost high summer, and I won't deny there is a singing out there in the corn patch that is more seductive to me than any big city Broadway show or cruise ship port-of-call. Thing is, I'm not sure what message it leaves me with. It goes well beyond "the world in a grain of sand" epithet. Most reflective people I know realize that the grandest, most profound truths in life do not need a grand stage upon which to reveal themselves . . . But if the pull of high summer in the corn patch is not revelation, then what is it? If not knowledge or beauty, then just which of truth's gifts is being offered?

I guess what it is, is satisfaction. Or to put it better, the contentment that comes from the experience of living within something in the same extent that it lives in turn in me. . . .

*I have stood on the high bluffs looking west over the
Pacific and on high bluffs looking east over the Atlantic, and
have tramped over a lot of land in between. I wouldn't deny
that all that has contributed toward making me a wiser, if not
exactly a better, person. But the exchange isn't there. I haven't
given any of myself to any of it, and none of it dwells in me
in return.*

The . . . "and yet" factor isn't there. It's in the corn patch.

Marilyn, the oldest of the sisters, nearly my age, practically
raised her younger sisters—the fate of the oldest girl in a large
family. She and husband Dennis were the first to build a new
house on the home place. She worked on the farm from childhood
as Dad's "right-hand man," especially since I, the oldest, left home
for awhile at an early age. "I loved everything about it," she says.
"I loved working with Daddy. I hated school. I would much rather
be home farming and fishing and hunting arrowheads, picking up
hickory nuts, working in the garden and taking care of my baby
sisters than any work or recreation anywhere. I worked for a doctor
in town after high school, but I would much rather have stayed
home and farmed. When Dennis and I decided to get married, he
was in the Army and in Germany and he begged me to come over
there with him. I was crazy about him, really wanted to be with
him, but I just could not bear to go that far away from home. I just
couldn't do it."

"You ask why we all stayed here," she continued, "or came
back like you, and I think the answer is simple. We like it here,
and we like each other. Why would we want to be anywhere else?
Our home life was not always happiness and joy, but I think that's
true of everyone's home life. Some acquaintances uptown have
criticized us for being too scared to leave home, but you can also
be too scared to stay home. As I grew older and found out more
about other families, I don't think our bad times were as bad as in

most families." Then she repeated: "We just like it here and like each other. I love our whole community and neighborhood, warts and all. I love our little town. Whenever I go there, I see old friends and acquaintances to gab with. We all care about each other and help each other when we can. There is nothing nicer than a little town and the countryside around it."

For myself, the only thing I can add is that I was willing to give up a secure financial future, if there is such a thing, to come home because of a deep yearning for tranquility. Growing up on a farm, I knew so many moments of utter peace when I listened to cows munching hay at night, or watched the sun come up out of the fog when I was already in the field, or sitting by the creek under a shade tree for a noon break watching the water flow by, or listening to the corn grow on a July night, or the sound of rain falling on the roof after dry weather. No matter how many times I walk or work the same land, there is always something new to discover and turn into even more tranquility. That is my reward, my destiny, my life.

CHAPTER 22

If Michelangelo
Had to Drive to Work

The London *Times* in the 1890s was prophesying that in fifty more years the streets of London would be covered ten feet deep or so in horse manure. Similar editorial predictions were voiced in New York City. Literary wisdom in those days declared that there had not been much of a problem with road apples in cities when only the richer people could afford to keep horses there. But with the rising tide of the Industrial Revolution, poorer people could also get the wherewithal to ride rather than walk, and that's when manure on the roads became a problem. I could not help but be reminded of editorial opinions today. Just as progress made it possible for poorer people to keep horses in cities way back when, so today poorer people can afford to buy cars, electricity, heating fuel, and hamburgers, especially now in China and other developing countries. One can discern just the faintest whisper of irritation in both yesterday's literature of city horse manure and today's reports of energy consumption. Things wouldn't be so bad if it weren't for all those poor folks getting uppity. It's okay for the Very Important People to own as many horses and carriages as they please, or, today, to burn up as much carbon as they deem necessary while they are attending to the world's business or heating their castle-like homes. It's those lower classes causing the problem.

The overweening notion implicit in the problem—that we were saved from drowning in manure by the invention of the everlasting oil well and the piston engine—is not really true. Human beings, not being idiots (even if we act that way sometimes), would have realized, if oil were not available, that manure, including the human kind, was valuable as fertilizer and could be turned into an industry, saving the nation from having to mine phosphorus and potash in other countries to keep our corn growing tall. If horses had remained a chief source of motive power, humans might have also wised up enough to stop cramming more and more people into cities, stacking them up in tall apartments where life would be impossible without excessive amounts of electric power. Society, at least the poorer part of it, would have invented garden farming sooner, where each residence, instead of having a three-car garage, would have three acres for a horse and their own food gardens, using the manure to fertilize the soil. This could have resulted in a landscape of small businesses, small farms, and small towns, all interacting in a vital and resilient local economy. It would have meant the flowering of garden farms much sooner rather than the flowering of automobiles and jets.

But, I hear you say, without automobiles and jets, there would be no progress in science, medicine, education, and so on and so forth. Depends on what progress means to you. We had already invented electricity before we invented cars and airplanes and nuclear bombs and drones. In fact we had electric cars before we had gasoline cars. Telephones and telegraph wires were humming away before the first dirigible blew up. We could do more of our traveling on the internet highway. There would be no enormous pressure to build concrete trails crisscrossing each other insanely from sea to polluted sea in an effort to connect every fast food outlet to every other, representing enough carbon emission to roast the moon like a Thanksgiving turkey.

Those who can't be happy without traveling could be riding trains and sailboats. Or stagecoaches. Chaucer and Shakespeare rode stagecoaches, and it seems to have improved their writing. Somehow the Bible got written and transported around the world, for better or worse, without the help of one single drop of gasoline. If Michelangelo and Beethoven had driven cars to work, I wonder if they would have had the time and concentration to become great artists. How much more great art did they produce because they didn't have to spend so much time waiting for lights to turn green, traffic jams to dissipate, and planes to get back on schedule? Andrew Wyeth said on many occasions that the best road to great art is to stay home.

Being able to fly around the world in a few hours has not deepened most travelers' understanding of anything except how to find cheaper tickets using the internet. A traveler who stays in a hotel for two weeks every year doesn't really get to know the people who live there. Unending travel has produced a society that acts like a bunch of bumblebees flitting from one clover blossom to another, never satisfied, never getting enough, never "finding myself" as so many travelers claim to be doing. A visit to the crumbling Parthenon today, Yankee Stadium tomorrow, and next week a cruise ship going up some river someplace. Meanwhile, travel has allowed countries with airplanes to drop bombs with impunity on countries that don't have airplanes. In our so-called advanced world, Michelangelo might have been killed by falling bombs before he taught himself how to carve his David.

When I apply all this to farming, practical people say that we can't go back to the way Grandfather did it. I say we can farm like Grandfather did, just as we can sculpt like Michelangelo did. Some of the most financially secure farmers I know are Amish, farming much like Grandfather did. I am tired of practical people telling me how impractical I am about farming when all around me I see examples of happy, successful people being "impractical" at farming. I watched a scene the other day so impractical I could not keep from laughing. Down the road where I was visiting came a one-row corn picker with an idled motor on it, driven by an

Amish farmer and pulled by six draft horses. Behind it came the farmer's son, on a tractor, pulling a wagon-load of corn that the picker had harvested. Impractical? Yes, but the Amishman had his reasons. He believes that using tractors on the road is okay, but not in field operations. The latter would tempt him to expand, to buy more land, and run his neighbor out of business. So he uses a motor-powered corn picker in the field pulled by six horses and pulls the wagon of corn down the road with a tractor.

Moving equipment on the road underlines a problem that is endemic to big-scale farming, and, as this Amishman shows, affects all farming. The more "modern" a farm gets, the more it must use the public roads to move machinery and haul in supplies and haul out stuff to market. In "modern" farming right now, it often takes longer to dismantle the huge cultivating and planting rigs for road travel and move them to the next farm ten miles down the road than it does to get the farms planted. Then, in the fall, it has become more "efficient" to haul the grain away in semitrucks. Because the country roads are single-lane with very little ditch space, these trucks get parked at field edges, where they often pack the soil into concrete. Because of this compaction, the concerned farmer won't drive the trucks over the field to where the harvester sits waiting to unload. Instead, he buys huge, expensive "carts" that he runs back and forth from combine to truck. These vehicles have huge tires, which are not supposed to compact the soil so much. Point is, the bigger the farms get, which is to say the farther apart the parcels of land the farmer owns, the more the cost goes up, and the less cost-effective is the operation. One could refer to the current large operation as "road farming" rather than field farming. An unsung advantage of smaller-scale, at-home farming is that it doesn't have to have as much gross income to pay for all this gross expenditure of time and energy. And is it really practical to own a million-dollar machine that you are going to use only two months or so every year?

It tempts the humorist to wonder about a future where farm machinery continues to grow larger while country roads do not grow at all. At some point dismantling the machine to fit the road will take as long as it did to put it together in the factory. Ah, I see the answer. Giant drones will transport giant machines from one parcel to another. Wonder what that will cost.

If a farmer believes that saving money is an essential part of making money, farming in one, at-home place is a more efficient profit center than the scattered parcels of road farming. Let us say a beef steer eats about thirty pounds of dry matter a day to reach market weight of 1,600 pounds. (This is just an approximation—exact figures vary all over the place, but most farmers figure about three pounds of feed produces a pound of weight gain.) The smaller farmer pays the transportation costs of hauling the animal or the meat to the market, but not the transportation cost of bringing in the grain and hay to the cattle lot or confinement building. He grows his own grain and hay at home. If the animal is processed on the farm, the cost of hauling the meat out is less than hauling out live animals, which cuts transportation costs even further. Whereas, when the hay and grain come from very far away, the transportation costs are enormous. Transporting hay is the greater cost, not only because of the greater bulk, but because that organic matter is not being returned to the soil where it grew. Yet road farming (or perhaps we should call it "travel farming") is at full tilt everywhere. As I wrote earlier, Saudi Arabia, having pumped its own deep wells dry, has purchased irrigated land in our American West to grow alfalfa for its dairy farms.

The amount of money saved with "at-home" farming grows much larger if we consider the impact of carbon dioxide emissions on the environment from all this road and travel farming. At-home farming saves lots of time, too. On the occasions when I have had to travel in city traffic, the thought always occurs to me that people who must commute into cities to work spend about as much time in their lives just waiting for traffic lights to change as it takes me to write a book. At-home farming requires no commuter time beyond walking to the barn in the morning. The blessings coming from that advantage are beyond money profits, and yet, as a society, we rarely take them into account, and in fact, think of being a "travel society" as an advantage.

In my neck of the woods we joke about an advantage that "travel farmers" have over the stay-at-home variety. Widely scattered showers are the rule here in August when rain is most needed and most profitable. If you have one farm on one side of the county and another on the other side, you can usually figure that the scattered shower will fall on one of them at least. That supposedly will mean enough increased income to pay for moving equipment back and forth between the farms.

Many of the problems facing farming today are not really about weather or climate change so much as about money gambling that encourages foolhardy farming. Farmers thought they might get rich growing corn for ethanol, so they plowed up hills and prairies that should never be submitted to annual cultivation. The resulting angst is so severe that it makes newspaper headlines and tempts people to think weather problems are greater today, when it is really economic greed and ignorance doing the mischief.

My old neighbor, now gone, always kept a goodly supply of surplus hay in his barn for emergencies. The emergency might come only once every ten years or so, but he was ready, and it took a lot of the worry out of contrary weather and climate change. "Extra hay in the barn is better security than money in the bank," he liked to say. "It also gives me the chance to sell that hay at a very high price when there is an emergency somewhere else and then replenish it with cheap hay in years of plenty."

As long as we had a variety of farm animals and a variety of crops on our farm, we could cope with adverse weather fairly well. But the year Dad and I listened to the fervid preachments about how specialization would mean bigger profits and got rid of all the livestock in favor of more cash crops, we put ourselves much more at risk. If weather decreased crop yields to below a profitable level, we had no livestock to eat the unprofitable yields so that we could eke out enough income from the animals to make it to a better year. Without hay or pastures in rotation for the animals, the grain crops needed more fertilizer, were more susceptible to drought, and weeds became much more problematic. When we then changed to a large dairy operation, the risk of having all our eggs in one basket, or, in this case, all our milk in one tank, was overwhelming. If milk prices were low, we had no other source of income to fall back on. In a drought year, we spent what profit there was from the milk on buying hay. But at least animals were a better risk than cash grain—hailstorms destroy corn, not cows.

Farming with the least risk means having a wide diversity of crops and livestock: chickens, hogs, cows, goats, corn, oats, wheat, barley, rye, hay, pasture, an orchard, a garden, and a pond full of fish. If one commodity fails, there are others to fall back on. This is why good Amish farmers run profitable operations, though small and using horses for traction power. And if you squeeze all the known facts out of Big Data, they will tell you the same thing.

CHAPTER 23

⟳

A Fable About the End of "Get Big or Get Out"

I n the early years of the 1970s when the big farming fever was running hot, I started writing about Marvelous Marvin Grabacre in the pages of *Farm Journal* magazine. He was just getting started in big farming and understood the economics involved precisely. The only way to get ahead was to keep getting bigger. Obviously, there would be only one farmer left and that would be Marvelous Marv. The economic system demanded it. Get big or get out. Eventually he owned all the farmland that a big tractor could fit on east of the Mississippi and then sorrowfully realizing that not even one-half of the nation's farmland was "a viable economic unit" he bought out his last competitor west of the Mississippi. "Get big or get out," he intoned, and that became the battle cry of agriculture well into the twenty-first century. He had his sights set on buying part of China next, or maybe Crimea and part of Ukraine, figuring he could sell off California, Arizona, and New Mexico to get the cash equity in his stock portfolio that he needed to attract that kind of money. Those states would soon run out of water anyway, he figured, so why not get rid of them while the price was still high.

But in 2018, while he was fuming because no one had a computer fast enough to suit him, he had a heart attack and died. Marvin Jr. took over, but he soon began wondering if his father's game plan was going to continue to work. In the Ukraine, for example, some of the investors in five-hundred thousand-acre farms there were selling their stock as fast as they could. Vladimir Megamoneylevich, an admirer of Grabacre, was finding that half of Russia was just not a viable economic unit for a farm and had purchased Poland, which was selling cheap right then. Megamoneylevich, like Grabacre, was known for his technique of making bigger fields out of big fields with tricks like running rivers, the Volga for instance, underground so he could farm right over them. When in an interview he was asked why he had purchased Poland, he replied: "Uberminsk ta loudervichnikoff," which roughly translated, means "get big or get out."

But Megamoneylevich was in trouble, too, like other monster farms. In Ukraine, they were declaring bankruptcy right and left, and China's big push to move all the people off the land and into cities to make big farms was not working out very well.

If you think this is only humorous absurdity and not also factual absurdity, just go online at farmlandgrab.net and see for yourself that this paragraph is not made up. Black Earth Farming, all 1,200 square miles of it, reported a loss of $26 million in 2008 and $39 million in 2009. More recently, it unveiled a loss of $18 million in 2013 compared to earnings of $7 million in 2012, the only profitable year in the company's nine-year history. Mikkail Orlov, after having sold off his Russian farming operations and shifted focus to Zambia, came back with a 4,888-cow dairy in Chechnya. Black Earth Farming sold twenty-eight thousand hectares of farmland in Russia to a company owned by Ukrainian oligarch Oleg Bakhmalyuk and US grain trader Cargill. Kinnevik, the investment group that was the biggest investor in Black Earth Farming, revealed that it was selling its Polish farm to cash in on gains in land prices and bankroll other opportunities in emerging markets because it couldn't make a $32.6 million interest payment.

Back to Grabacre in 2018, Marvin Jr. saw all this and knew he had to do something to revamp his bottom line. But what? That's when he got his great idea. By 2020, people all over were clamoring for land of their own, and it occurred to him that, if he could sell off his acreage by re-packaging it into small parcels of two- to twenty-acre mini-farms, he could make a fortune. He tried to multiply his nine-hundred million acres of American farmland by twelve thousand dollars per acre but could not keep track of all the zeros. For sure, he would be the world's first multi-trillionaire. He started un-grabbing.

And so it came to pass. That old Bible had it right. The meek inherited the Earth and a period of peace and plenty occurred like nothing ever seen before. Marvin Jr. did save Iowa for himself, however, and found that it was small enough so that he could actually turn a profit some years. He would smile then and say, "Get small or get out."

CHAPTER 24

The Real Background Behind the Fading of Industrial Farming

F arming has become sort of like a computer game, ignoring the wisdom of tried-and-true traditions. A friend, with his eye on the Chicago "Bored" of Trade, stunned our neighborhood by growing only soybeans one year. To ignore corn in the Corn Belt is like embracing atheism in the Vatican. He guessed right—the first time. In his second year of denying the existence of corn, the bean market was oversupplied and his bottom line was appropriately named because it sank to the bottom. Another neighbor decided all corn was the way to go and that year drought took away his potential profits. If he had planted his usual acreage of wheat alongside the corn, it would have made up for what he lost. Wheat can stand dry weather better than corn.

The whole gamble-farming gambit that caused the rocketing rise in farmland prices between 2008 and 2014 was fueled by investment companies looking for protection after the stock market crash in 2006–07. Farming from 2008 to 2011 made a better return on investment than even gas, oil, retail marketing, or telecommunications. (See the Economic Research Service online at www.ers.usda.gov if you doubt me.) The philosophy behind this investment rush was the ancient observation that land would always be there even if the paper money disappeared into the cloudy atmosphere of greed. As the economists of the "New Now" philosophy are saying, in the future not even the smartest wheeler-dealer in derivatives (or degeneratives, as I prefer to call them) is going to be able to squeeze fortunes out of paper. What wealth will be available will have to come from, or be supported directly by, things of real value, like good food. The hedge fund gangs have used up all of nature's slush funds. Buy Poland. The vodka is worth it.

Headlong investment in farmland, thereby running up its paper money value, has never worked for long. As recently as the 1980s, after insurance companies and retirement funds and wealthy individuals invested heavily in farmland, the market collapsed. Remember? It wasn't that long ago. The same players did it again in the run-up to 2015's collapse.

The best history lesson in monster farming came, believe it or not, in the horse-powered era. The greed dreams then were called "bonanza farms," and they were especially prevalent in Iowa, North and South Dakota, and western Minnesota, where plenty of cheap land was available. Here farms of unbelievable size, at least at that time—twenty-five to fifty thousand acres, even seventy-five thousand acres—came into existence, each manned by hundreds of humans and horses, bossed by several layers of management, and funded by wealthy investors. As it turned out, about the only money "made" in these enterprises came from the rise in land values. Much of the land had been bought for little more than a dollar an acre. If nothing else, the very act of wheeling and dealing in stocks that were supposed to represent real dirt acres, made the price continue to rise independently of the realities of nature or the land's ability to grow crops at a profit.

The huge bonanza holdings lasted from about 1880 to 1910, when sanity and logic, along with rising labor and land costs, caught up with greed. But in their heyday, these farms looked uncannily like the huge farms we have today, except that they were powered by horses not tractors. Look at an aerial view of those much-heralded gigantic grain farms of Brazil today, with battalions of combines stair-stepped across seemingly endless acres. In your mind, put teams of horses and binders in place of the combines, and you have almost the exact landscape of a Dakota bonanza farm in 1885.

The sheer vastnesses of these old farming empires as depicted in the photos in Hiram Drache's book, *The Day of the Bonanza,* are awesome to behold. A thousand-acre field of hand-shocked wheat always makes me shiver, since I have shocked wheat by hand. A line of bundle wagons pulled by teams of horses stretches completely across the horizon of another picture; I can count twenty of them, but the line stretches beyond the picture frame. Another photo shows forty-three binders cutting wheat, side by side. And remember that in most cases, the laborers were non-residents. Both they and the horses had to be housed and fed on the farms. When not in use, the horses often were herded to high mountain pastures to graze and the human workers returned to the streets of towns and cities from which they had been scraped. The logistics of this kind of operation are on a par with moving an army into battle. Some of the bonanza farms hired fifty to a hundred men and stocked six to eight hundred horses. To anyone, like myself, who has lived horse-farming days, it is almost beyond imagination. It is vexing enough to have to deal with one team and one hired man. Surely no one with any experience could think that this scale of farming would work for long. Obviously, most of the stockholders, from faraway wealthy families, didn't know a gang plow from a grain binder. But the land went up in value every year, so their money stocks went up in value, too, and it looked like a good deal. Among many other lessons to be learned, the Amish notion that farming with horses removes the temptation to increase farm size is another of those beliefs that ain't necessarily so. When one is farming with paper-purchased horses, ain't nothin' necessarily so.

Remember also that this was often Red River Valley country, where enormously unpredictable weather should have sounded a cautionary note for investors. One of the significant footnotes to this sad history is recorded on the 1898 two-cent postage stamp commemorating the Omaha Exhibition. The stamp shows eighteen gang plows being pulled by horses in a boundaryless field, taken from a photo of real life on the Amenia and Sharon Farm, one of the biggest of the bonanza farms. What neither photo nor stamp can tell you, but which Mr. Drache does in his book, is that all those horses, gang plows, and day laborers were busily turning under a *hailed-out wheat field*. This is the farming fact of life that no investment banker, however brilliant, no warped board trader, however much a risk-taker, can escape with pieces of paper.

The way that weather always sways uncontrollably over farming to make it exceedingly risky for outside investors was not lost on Mr. Drache, writing three-quarters of a century later. I once visited him because I was overwhelmed with curiosity about a historian and college professor who was also a real commercial farmer. His farm was in the Red River Valley he wrote about, so he knew the lay of the land, figuratively and literally. I stood in his barnyard and looked out maybe a mile over the level landscape. It was very hot and dry that day, and the corn looked poorly. "You might find it hard to believe," he said matter-of-factly, "but right where you are standing you could be in five feet of water in spring flood times."

The financial side of these bonanza farms is well documented by Mr. Drache and others and is most interesting to study. There were constant disputes between shareholders and management. Also, constant arguments flashed back and forth by telegram about whether the reported outgo and income were accurate. Some years, some bonanzas actually made a profit, but mostly when all operating expenses were included, there was a net loss compensated for awhile by the increase in land values. So what's new?

I have in other books quoted the words of William Dalrymple, part-owner of one of the biggest and most famous bonanzas, when his family sold out in 1917: "My brother and I have decided to give up operating the farm and divide it into small farms. . . . Big farms were once good for publicity. But economic conditions have changed. . . . It will be better for the state, for the towns and cities . . . to have a great many small farms in place of one big farm."

Perhaps a better quote comes from someone who was actually there, in Iowa, when these big farms were starting up, already voicing his apprehension. Prince Kropotkin (mentioned in chapter 17, "Have We Deflowered Our Virgin Soils?"), an economist hailing from Russia and residing in England when not traveling, wrote what I think is one of the most revealing economic studies of his time, and that is still applicable today: *Fields, Factories, and Workshops* (published in 1907). Kropotkin was a socialist, and although his book was highly praised by all sides, even in the United States, American society as a whole, steeped in fear of socialism, paid it little heed. After traveling in Iowa in 1878, and observing the first mammoth wheat farms failing and being broken up into small farms, he wrote:

> *In fact, over and over again it was pointed out, by . . . many other writers, that the force of "American competition" [with Europe] is not in its mammoth farms, but in its countless small farms upon which wheat is grown in the same way as it is grown in Europe. . . . However, it was only after I had myself a tour in the prairies of Manitoba . . . that I could realize the full truth of the just-mentioned views. . . . [E]ven under a system of keen competition, the middle-sized farm has completely beaten the old mammoth farm, and that it is not manufacturing wheat on a grand scale which pays best. It is also most interesting to note that thousands and thousands of farmers produce mountains of wheat in the Canadian province of Toronto and in the Eastern [American] States, although the land is not prairieland at all, and the farms are, as a rule, small.*

The force of the "American competition" is thus not in the possibility of having hundreds of acres of wheat in one block. It lies in the ownership of the land, in a system of culture which is appropriate for the character of the country, in a widely developed spirit of association, and finally, in a number of institutions and customs intended to lift the agriculturist and his profession to a high level which is unknown in Europe.

Sounds as much like capitalism as socialism to me, but in either case, it stands as a very early observation about the weaknesses of big farming: No matter how much technology comes along to make mammoth farms seem practical, they can't be sustained but must keep growing until they collapse. The small, intensive, mostly organic grain/livestock farm perfected in the eastern United States during the early twentieth century is still the most economic way yet devised to produce the meat, potatoes, gravy, and bread that most of us crave.

Big farm technology may be pricing even so-called cheap factory food out of the poor man's market. In recent years, beef prices have climbed beyond affordability for poorer people. Numbers vary from season to season, but a neighbor told me in 2015 that his annual fertilizer bill alone was upwards of $80 an acre, or $400,000 on a five thousand-acre operation. Chemical sprays at $20 an acre added another $100,000—probably more than that, especially if unforeseen insect problems break out and extra sprays are needed. If the farmer was buying high-priced genetically modified seed corn, that could have added at least another $80 per acre. If the farmer was renting land for $200 an acre—could be more—and half of his land was rented, there goes another half million dollars. If he owned the farm or the equipment on borrowed money, he could have been paying a frightening amount of money interest. And there are land taxes to be paid also, which have been going up. I don't know how to figure the cost for fuel and machinery since fuel prices have recently come down and the cost of machinery has gone up. Tractors and harvesters cost around $500,000 right now, and the price continues rising. There are harvesters on the drawing board projected to sell for a million dollars each. All in all, the farmer has somewhere around two million dollars invested in five thousand acres of corn before he harvests a single bushel. Then he must not only harvest it but haul it to market or to his own storage bins, and pay drying costs. If the fall is wet and the crop a little late, profit or loss can depend on how much LP gas or natural gas he uses. Since corn must be dried down to 15 percent moisture for safe storage, the cost of doing that on really wet corn can be as much as 10 cents per percentage point, or $1.50 a bushel. Even at 5 cents a point on corn coming out of the field at a more normal 20 percent moisture or with the current cheaper prices, it means 25 cents a bushel—no small cost on a couple million bushels. One recent year, a significant amount of corn in storage, even after drying, was attacked by molds like vomitoxin and could not be used for livestock feed. No wonder crop insurance has become so important.

The cost of corn is not borne only by the farmer. Elevators have their headaches, too, with mountains of corn stored outside in the weather until it can be moved. The trains, barges, and trucks have to make a little money, too. Then the ultimate madness: The corn is shipped to animal factories often hundreds of miles away, and then, after it is turned into meat, milk, and eggs, much of that is shipped back to where the grain came from.

The reason farmers are willing to take such a risky course is that, if everything goes well, there's still a chance of making a lot of money on really good land. And if they fail, the government will save them with subsidized insurance. Five thousand acres of corn, if it averaged two hundred bushels per acre and could be sold at five dollars a bushel, is a heap of money. A five million dollar heap. Even if your costs are four million dollars, you still net a million.

But how rarely does the corn farmer lock in a five dollar a bushel price and harvest two hundred bushels per acre? Sometimes on the best ground and under the best management, the yield might be even higher, but more than likely it will average around 170 bushels per acre (although a new world yield record of over five hundred bushels per acre has been achieved, whetting every farmer's appetite to keep on going). The price of corn has been stumbling around $3.80 a bushel (as I write in 2016), which isn't enough to break even, say economists. The point is rarely made, but if you don't break even, a great two hundred bushel per acre yield is a bigger loss than a one hundred bushel per acre yield. And at the moment you can't lock in a higher price on the futures market because the futures price is even lower. No wonder that right now, in 2016, sales of big tractors are falling dramatically, while the sales of small tractors to garden farms are rising dramatically.

What is going to happen? If we only knew. But the trend is toward more small market garden farms. It is not outlandish to predict that in the future food will come more from this source because smaller-scale farmers can produce food at lower costs, especially if they own their land and use older or simpler equip-ment or even horses, supplemented by their own labor. With their own open-pollinated seed corn or low-priced traditional hybrids, their own barn manure for fertilizer, their own "old-fashioned" slatted cribs for low-cost drying instead of artificial heat, they will be able to produce, say, forty acres of corn at a cost of $2 a bushel. That leaves them a profit of $2 a bushel on $4 corn. Forty acres worth, at a realistic yield of 120 bushels per acre is 4,800 bushels or a profit of $9,600—not bad part-time wages.

I say this not as a critic of industrial grain farmers, some of whom are close friends, close neighbors, close family members. I sympathize with them. Neither they nor anyone else could have foreseen what would happen when farming became driven by the madness of making nature bow to money rather than the other way around. It could mean that today's larger-scale farmers are the last of their kind or will be limited to crops grown for industrial products or food additives, in which the growers will be little more than hired hands of industrialism.

Just how much further can we go in the pursuit of bigness? Finally, when the whole world is one farm, as Marvelous Marv Grabacre envisaged, and still does not return a profitable investment in paper money, will there be big money talk about rocketing off to find another planet to fill up with corn, wheat, and soybeans? Won't it finally dawn on everyone that it would be more sensible to let the world fill up with small farms operated by garden farmers who want only enough money to keep on doing comfortably what they love to do?

Chapter 25

In Praise of
Rural Simplicity
(Whatever That Is)

Nothing irritates a contrary farmer more than getting stereotyped. So I was a bit disturbed while reading an interesting article about attention-deficit/hyperactivity disorder (just getting those letters all out in correct order is enough to give me ADHD) suggesting that the condition is not really a bodily or mental affliction but a natural state for some people, especially children, brought on by the over-regulated, proscriptive world we live in ("A Natural Fix For A.D.H.D." by Richard A. Friedman in the *New York Times* editorial section, Nov. 2, 2014.) All fine, as far as I know. What made me gnash my teeth once again was the example that the learned scientists used. They suggested that ADHD people would be right at home in a hunting and gathering society, like in Paleolithic times, when daily life shifted rapidly from one exciting, dangerous situation to another. It was not until humans settled into the boring routine of sedentary agriculture, said the scientists, that they became estranged and out of touch with the rest of society and started suffering from what would later be diagnosed as ADHD.

Even before I object mightily to the notion that farming is boring, I think the learned scientists have it all wrong about hunting and gathering, too. Since none of us has ever lived in Paleolithic times, we can say anything we feel like saying on the subject and get away with it, but I have done my share of hunting and gathering and take strong exception to the article's claim that "a short attention span was useful to hunter-gatherers." Quite the opposite. Successful hunters learn to discipline themselves to long hours of sitting quietly in tree stands or in blinds or on stumps waiting patiently for their quarry to come within range. Learning how to track prey and then how to shoot straight both require much discipline. One of the problems with hunting today is that too many would-be hunters do have short attention spans and rarely kill anything except other hunters. They love to roar up and down country roads in pickups, hooting and hollering and blasting away in all directions, scaring the game into the next county. The traditional English fox hunter, galloping across the fields on horseback while tally-hoing was the kind that qualified for ADHD diagnosis.

As for farming being boring, please. I grant that there are days when you might spend hours in a tractor cab, listening to talk radio rants or gabbing on your smart phone while the tractor drives itself. But the second you quit paying attention to what's going on, or almost drop off to sleep in boredom, bells and whistles are likely to start clamoring away, indicating a loose belt or a broken pin or a plugged up auger or the embarrassing fact that you just plowed halfway through the township dirt road bordering your field. If anything, ADHD-afflicted people should by all means get into farming. They are likely to succeed like no other personality type.

Every day on the farm is full of gut-wrenching situations. Farmers live with one eye on market reports streaming across the computer screen and the other on the sky, scanning hopefully for either fair weather or rain. He or she must be ready at all times to click a computer button on the futures market reports that might mean losing or gaining a hundred thousand dollars or so. Or if you are a new age farmer trying to grow for the local market or for CSA sales, you must be on your toes constantly to be in close touch with your customers' moods.

And the tension never ends. Used to be there would be time in winter for a little relaxation. But the government keeps making regulations in all seasons, and the connivers keep finding ways to get around the regulations, so one must be ever vigilant and ready to outwit both. Nor do CSA customers quit wanting more lettuce just because the weather has turned cold. What farmers yearn for more than anything is a month or two of boredom every year. Or they need medicine that will make sure they never lose their ADHD tendencies.

Where stereotyping ends and bigotry begins is hard to determine, especially when the subject is farming. I have been the target of both, and the result, as far as my feelings are concerned, is the same. I should be used to it by now, after being made fun of as a clodhopper in childhood, and dumbfarmer (one word) as a teenager, and not so long ago, being told by university intelligentsia that I should, as a writer, "stick to subjects like corn and leave more critical issues to those of us who are better informed." But it still rankles me, especially since I thought surely that prejudice against farmers had passed into history, at least among the "better informed." But recently I saw a quote from a French philosopher, Pascal Bruckner, in *The Atlantic* ("How to Talk About Climate Change So People Will Listen" by Charles C. Mann, September 2014) criticizing those of us, especially environmental writer and honorary contrary farmer Bill McKibben, who are suggesting that an answer to many of our woes, including climate change, would be a nation of careful, small-scale farmers. Bruckner thinks of this as wanting to go back to a "puritanical straitjacket of rural simplicity." Perhaps I am too sensitive, but to me that statement reeks of bigotry against farmers.

I tend to agree with philosopher Bruckner on a number of things (in his book *The Fanaticism of the Apocalypse*), but, oh my. If our "educated classes" think that careful, small-scale farming is a "straitjacket of simplicity," I wonder exceedingly if there is any use in trying to straighten out their jackets, much less their minds. I like the idea of rural simplicity, but after eighty years of trying to achieve it, all I can say is that rural life is a whole lot more complicated than life in the ivory towers. I know, because I have been there, too. Perhaps if I can express the knowledge I've learned, about corn for instance, in the more rarified language of philosophy, I might be able to change the kind of mindset that Mr. Bruckner is displaying here. Since I have a degree in philosophy, too, maybe I should start talking like it, suggesting far-out notions about what is wrong with society in high-sounding words, coining new ones as I go along—like philosophers and economists do.

The biggest challenge confronting us right now in my opinion is not climate changing but money changing. We are in the throes of berserkonomics, or, in dumbfarmer terms, stalkrotnomics. We have forgotten that cornstalks can't adjust their growth to the ups and downs of the stock market or grain market gamblenomics. When corn is eight dollars a bushel one year and four dollars the next, we are under the sway of Keynesiastic-Friedmanic Oscillation and there is nothing simplistic about it. The goofiness of money can be expressed in the world of rural simplicity by applying KFO as influenced by FRLIRWT (Federal Reserve Low-Interest-Rate Wishful Thinking) and HGF (Human Greed Factor) to corn production.

Allow me to do a little philosophical oscillation here to show what I mean and how complex rural simplicity can be. Sometimes more "profit" can come from turning the whole "get big or get out" kind of farming upside down—or oscillating it, so to speak. By my own puritanical method of calculation, I figure I once made an unbelievable $550 net profit on a per acre basis from my measly half-acre of corn, while modern Big Data farmers with five thousand acres of corn lost on average eight dollars an acre. My half-acre yielded fifty bushels of corn or one hundred bushels per acre—almost a crop failure by Big Data numbers. The corn price was right around $4 a bushel at that time, so my fifty bushels, on a per acre basis, grossed $400. Of course I can only assume that yield hypothetically because maybe deer would have eaten the other half-acre. Many big-time commercial farmers, no longer in the straightjacket of rural simplicity, were getting about two hundred bushels per acre, so it would seem that they were way ahead of me in terms of "good" farming as measured by money. But by using legume rotations and my own barn manure, my out-of-pocket fertilizer cost was almost zero. My out-of-pocket herbicide cost was zero because I didn't use any. My out-of-pocket seed cost was also zero since I save my own seed. My out-of-pocket land cost was zero since we have the land to enjoy as part of our homestead whether we grow anything on it or not. Harvest labor cost was zero since we husked the corn by hand as part of a family day of fun. Drying costs were virtually zero because I dried the corn with free, natural air in a slatted crib I built myself years ago for insignificant out-of-pocket costs. I didn't have to mechanically shell the corn, so that was another savings—I fed it to my animals on the cob. Hauling cost involved driving the pickup about eight hundred feet from field to corncrib. I did have some cost in using a rotary tiller to prepare the soil. I suppose I should charge a labor cost for that, too, although I consider my corn a recreational activity like playing golf, only a whole lot cheaper. Let us say I had real out-of-pocket costs of about $50 on a per-acre basis. That leaves me a net profit of $350 per acre with $4 corn. But if I had bought the corn at our closest farm supply store, I'd have had to pay $6 a bushel instead of $4 because the elevator was charging that much for handling grain in small amounts. (If you want to know how wacky farming is, it is easier for an elevator to supply me with a truckload of corn than a sack full.) So I figure I saved another $2 a bushel by growing my own. I can say my profit after all costs was around $550 total per acre, which industrial corn farmers would die for. And what if I developed an improved strain of this corn by cross-pollination worth far more as seed than as feed. Just to complicate the matter further, I know a certified-organic, small-scale farmer (the ultimate in puritanical straightjacketing) who was getting $12 a bushel for his corn at that time, not $4. He doesn't get the big yields necessary to make a "profit," but without the expenses of commercial chemical farming, he doesn't need to. He told me one recent year that he "made so much money it was embarrassing." So much for rural simplicity.

But there is a kind of rural simplicity that really is simple, and we simpletons need to extol it like learned philosophers would do if they only knew what farm life was really like. Have you heard of the Law of Unequivocality? No, you haven't, because I just made that up. Remember, I have a degree in philosophy and am allowed to do that. One example of this law is often observed in the land of rural simplicity: If you stand out in the rain unprotected, you are going to get wet. Another example is that one plus one equals two. It's always true. If you have fifty bushels of corn that really didn't cost you anything much to grow because you'd rather do that than play golf, and corn at the elevator is going to cost you six dollars a bushel if you have to buy it, your fifty bushels are worth $300 to you—a simple, straightjacket way to apply the Law of Unequivocality. Meanwhile, in the bimbo world of gamblenomics, unshackled by the straightjacket of rural simplicity, the sum of one plus one depends on whether the head of the Federal Reserve clears her throat once or twice when discussing interest rates.

But the Law of Unequivocality isn't always enough to figure out the complex world of rural simplicity. Let us say that, in my puritanical straightjacket, I keep a cow whose milk I drink and whose cream I pour on my strawberries fresh from my garden, which I actually did for many years. The cow ate mostly grass and clover that also enriched the land I planted to corn in rotation with these forages. If I counted all the time I spent producing those dairy products at wages that a philosopher makes in universities now, I have a notion that I would have to price a quart of my milk at three or four times what store milk sells for. This is where the world of rural simplicity starts to get really complicated. The straightjacketed economist tells me I am foolish to spend all that time producing my own dairy foods when it would be cheaper just to buy them and go to work as a philosopher. Let me try to count the ways why this is berserkonomic thinking.

First of all, out here in my puritanical straightjacket, time is not money; time is life. And I love my life. I love it so much I made a decision to pursue it even though it meant we would not have much money to spend on stuff we didn't need. Second, I don't know how to put a true monetary value on this milk, cream, butter, cottage cheese, and all the baked and cooked foods that include these dairy products in their recipes because my kind taste so much better than the store-bought kind. The whipping cream you buy is just not anything compared to whipping cream made from ladling the cream off the top of a bucket of Guernsey milk.

So far I am talking merely about taste. Although the arguments rage on and on, there is mounting evidence that dairy products raised and consumed my way are more nutritious than what you buy in the supermarket. Moreover, saturated fats are good for you, say more and more doctors and scientists who reject the old cholesterol-phobic view that fats cause heart disease. Unpasteurized milk is also more nutritious, many experts say, if cows are healthy and good hygiene is followed in processing the milk. I don't care who is right. I love whole milk, cream, and butter. I have consumed them with gluttonous abandon all my life, and for years the milk was unpasteurized. My heart is in pretty good shape in my eighties. I still have all my teeth. If these foods really are more healthful than the commercial products, think how healthy we are in our puritanical straightjackets. What's that worth?

But the complexity inherent in this kind of rural simplicity does not end with the dairy products alone. Also being produced in close connection with the milk are meat, eggs, fruits, vegetables, nuts, herbs, wood for heat and lumber, jams, lard, hay, and pasture for the cows and chickens, all without much help from farm chemicals or fossil-fueled shipping costs. All these products come from simple activities that do not require commuting through traffic jams and are part of a very complicated whole. For example, when I am making hay I am also making nutritious, low-cost milk and meat. If I do all these activities correctly, I am simultaneously making the soil from which it all comes richer and more health-producing. When I try to figure out what an hour of my time is worth in money, to which of these varied activities should I ascribe it, since often I am doing more than one of them at the same time? This is especially true for me in my puritanical straightjacket because while I am making hay, milking cows, cutting wood, weeding gardens, or whatever, I am also in my head writing essays like this one. Other people in their puritanical straightjackets are composing music in their heads while they farm, or seeing landscapes they will render into art, or thinking up more interesting ways to present truth to students they are teaching, or keeping their bodies in good shape for sports or off-farm physical work they engage in to make a little extra money. If you had asked me years ago when I was milking cows by hand what advantage I gained in doing so, I would have bragged that hand-milking is an almost perfect exercise for developing arms that can swing a ball bat effectively. Physical therapists said it, and I could provide evidence of it because I was batting over .500 against pretty fair competition. Squeezing the milk out of the cow's teats builds formidable forearm muscles.

Accountants toting up farm production numbers have no way of adding in these kinds of profits. Nor is it evident, unless you live in rural simplicity, how much sports, America's most pervasive religion, and other recreational pastimes are woven into the warp and woof of rural simplicity. In the heyday of our hometown softball team, it came out as we all sat in a bar over after-game beer that every single player on our team had done his share of sweating in barns putting up hay. Nothing like putting up hay in 98° temperatures to harden your body for all-night tournaments.

Add horses to this puritanical straightjacket and the concept of time as money clouds over even more. Using horsepower instead of piston power on farms saves no telling how much money in fossil fuel and machinery expenses. You can't farm as many acres with horses as tractors perhaps, but you don't have to, as the bank account of almost any older Amish farmer will attest. But the point is that all the horse farmers I know love their horses, just as pet owners love their cats and dogs. Farming with horses is part of the enjoyment of their lifestyle.

Which brings up a whole vast range of other pleasant complexities to be enjoyed while living in the puritanical straight-jackets of rural simplicity. When time is not money, there can be more of it made available just for fun. There are, in fact, so many recreational opportunities available without leaving the farm that some of us don't do much traveling at all. Not traveling cuts down on carbon emissions, to say nothing of the cost of living. Some of us become amateur archeologists hunting Indian artifacts in our cornfields. Or birdwatchers. Horseback riding is not just a hobby for the rich but for any farmer who wants to do it. Before my mother got her own riding horse, she rode one of our workhorses around our farm, side-saddle. If you have hills on your farm, you have your own sledding and ski resort. Farm ponds mean fishing, swimming, and, in our neighborhood, hockey above all. Interestingly, back in the days when all farms in our neighborhood were small, fairly subsistent livestock and grain operations, we never had to think about scheduling activities for weekends because no one did the forty-hour week thing. Our way of life kept us in control of our time, not vice versa. Whenever the ice was ready for hockey, we were ready, too. We got the chores done early or late and then we flocked to the neighborhood pond to try to kill each other with hockey sticks. Didn't make any difference if it was Wednesday or Saturday.

Now, like good philosophers are supposed to do, let us stretch our minds to really wild musings. Suppose we had a nation of careful, small-scale, garden farmers—let's say 150 million of them, leaving another 150 million or so people to think up new philosophical and economic theories for things we don't understand. This would mean less carbon emissions, less soil compaction and erosion, less financial instability, less hunger, surely less joblessness, and maybe less war. It would even mean less traffic accidents because people would be so busy enjoying being puritanical (or not) at home in their rural simplicity that they wouldn't be tempted to go out in public and scream political diatribes at each other. They would be too interested in occupying their farms to want to occupy Wall Street.

Let's suppose a little more. Suppose all these little food-producing farms produced food as a spare-time hobby, as many already do. Can you imagine what a sea of change would take place in the mad world of money if so many people produced their food in the same spirit with which they now play golf, producing maybe a little more to keep the philosophers well fed? There might not even be a Wall Street to occupy because all its stockbrokers would be off enjoying rural simplicity.

Trying to visualize an economy not run by paper money would require more philosophers as well as more farmers. Society would need clever words and phrases to soothe those frightened by change. Maybe call the new rural order "Static Superconomy," with catchy slogans like "where everyone is a millionaire because no one is." Suddenly we would achieve the idyllic world of peace and plenty that the philosophers all yearn for. And if that got a little too boring for some, there would always be sports around to satiate the human craving for violence.

Right out there next to the Statue of Liberty, we could erect a statue of the garden farmer who made it all possible.

ABOUT THE AUTHOR

BEN BARNES

Over the course of his long life and career as a writer, farmer, and journalist, Gene Logsdon published more than two dozen books, both practical and philosophical, on all aspects of rural life and affairs. His nonfiction works include *Gene Everlasting*, *A Sanctuary of Trees*, and *Living at Nature's Pace*. He wrote a popular blog, *The Contrary Farmer*, as well as an award-winning column for the Carey, Ohio, *Progressor Times*. Gene was also a contributor to *Farming Magazine* and *The Draft Horse Journal*. He lived and farmed in Upper Sandusky, Ohio, where he died in 2016, a few weeks after finishing this book.

 **green press** INITIATIVE

Chelsea Green Publishing is committed to preserving ancient forests and natural resources. We elected to print this title on 100-percent postconsumer recycled paper, processed chlorine-free. As a result, for this printing, we have saved:

26 Trees (40' tall and 6-8" diameter)
12 Million BTUs of Total Energy
2,211 Pounds of Greenhouse Gases
11,992 Gallons of Wastewater
802 Pounds of Solid Waste

Chelsea Green Publishing made this paper choice because we and our printer, Thomson-Shore, Inc., are members of the Green Press Initiative, a nonprofit program dedicated to supporting authors, publishers, and suppliers in their efforts to reduce their use of fiber obtained from endangered forests. For more information, visit: www.greenpressinitiative.org.

Environmental impact estimates were made using the Environmental Defense Paper Calculator. For more information visit: www.papercalculator.org.

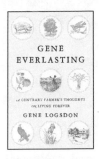
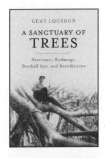
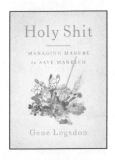
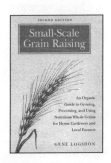

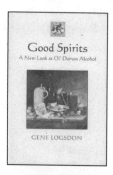
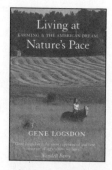
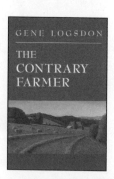